LOCOMOTION PAPERS                                          LP247

# THE
# BRISTOL TO PORTISHEAD BRANCH
## WITH THE
# BRISTOL HARBOUR RAILWAY AND CANON'S MARSH BRANCH

by

Colin G. Maggs

Broad gauge Bristol & Exeter Railway 4-4-0ST No. 72 at Portishead *circa* 1870.
*Author's collection*

# THE OAKWOOD PRESS

© Colin G. Maggs, 2020.

ISBN 978-0-85361-745-7

Printed by
Claro Print Ltd, Office 26/27, 1 Spiersbridge Way, Glasgow, G46 8NG

Junction signals at Parson Street 21st April, 1970: the Portishead branch curves to the right immediately beyond the bridge.  *Author*

*Front cover:* A dmu at the up platform, Clifton Bridge, 22nd August, 1964.  *Michael Farr*

*Back Cover:* Upper - A dmu near Oak Wood signal box, 30th April, 1964.  *Michael Farr*
Lower - Port of Bristol 0-6-0 diesel-mechanical Hudswell Clarke, No. 26 *Douglas* left and the same builder's 0-4-0 diesel-mechanical *Gordano* right, 14th February, 1972.
*Revd Alan Newman*

Published by
The Oakwood Press, 54-58 Mill Square, Catrine, KA5 6RD
Telephone: 01290 551122    Website: www.stenlake.co.uk

# Contents

| | | |
|---|---|---|
| Bibliography and Acknowledgements | | 4 |
| Introduction | | 5 |
| One | Early Schemes | 7 |
| Two | The Line Opens | 15 |
| Three | Portishead Pier & Shipping | 19 |
| Four | Portishead Dock | 27 |
| Five | The Railway Evolves | 31 |
| Six | The Line is Purchased by the Great Western Railway | 35 |
| Seven | Developments | 45 |
| Eight | The Freight-only Years | 53 |
| Nine | Description of the Line | 55 |
| Ten | Locomotives & Coaches | 117 |
| Eleven | Timetables and Train Working | 123 |
| Twelve | Signalling and Permanent Way | 131 |
| Thirteen | Regeneration | 133 |
| Fourteen | Proposed Re-opening to Passenger Traffic | 141 |
| Fifteen | The Bristol Harbour Railway & Wapping Wharf Branch | 143 |
| Sixteen | Description of the Bristol Harbour Railway | 151 |
| Seventeen | The Canon's Marsh Branch | 169 |
| Eighteen | Description of the Canon's Marsh Branch | 177 |

# Appendices

| | | |
|---|---|---|
| One | Traffic Dealt With At Branch Stations | 180 |
| Two | Industrial Locomotives | 184 |
| Three | Speed Restrictions | 187 |
| Four | Loads | 188 |
| Five | Signal Boxes | 190 |
| Six | Log of Runs | 191 |

Index ............ 192

# Bibliography

Clinker, C. R., *Clinker's Register of Closed Passenger Stations and Goods Depots* (Weston-super Mare: Avon-Anglia Publications, 1988)

Cooke, R. A., *Track Layout Diagrams of the Great Western Railway and BR Western Region, Section 19A* (Harwell: Author, 1992)

Cummings, J., *Railway Motor Buses and Bus Services in the British Isles 1902-1933* (Oxford: Oxford Publishing, 1978)

Fleming, D. J., *Raising the Echoes* (Cheltenham: Line One Publishing, 1984)

Fleming, D. J.,*Pannier Tanks and Shunting Poles*, Part 1 (Bristol: Author, no date)

Hateley, R., *Industrial Railways and Locomotives of South Western England* (Melton Mowbray: Industrial Railway Society, 2011)

Maggs, C. G., *A History of the Great Western Railway* (Stroud: Amberley, 2013)

Maggs, C. G., *Bristol Railway Panorama* (Bath: Millstream Books, 1990)

Maggs, C. G., *The Branch Lines of Somerset* (Stroud: Amberley, 2011)

Maggs, C. G., *The Bristol Port Railway & Pier* (Tarrant Hinton: Oakwood Press, 1975)

Maggs, C. G., *The Weston, Clevedon & Portishead Light Railway* (Oxford: Oakwood Press, 1990)

Marshall, J., *A Biographical Dictionary of Railway Engineers* (Newton Abbot: David & Charles, 1978)

McKenna, F., *The Railway Workers 1840-1970* (London: Faber & Faber, 1980)

Oakley, M., *Bristol Railway Stations 1840-2005* (Bristol: Redcliffe Press, 2006)

Oakley, M., *Somerset Railway Stations* (Wimborne: Dovecot Press, 2002)

Redwood, C., *The Weston, Clevedon and Portishead Railway* (Weston-super-Mare: Sequoia Publishing, 1981)

Robertson, K., *Great Western Railway Halts Volume One* (Pinner: Irwell Press, 1990)

Robertson, K., *Great Western Railway Halts Volume Two* (Bishop's Waltham: KRB Publications, 2002)

Simmons, J. & G. Biddle, *The Oxford Companion to British Railway History* (Oxford: Oxford University, 1997)

Smith, M., *The Railways of Bristol & Somerset* (Shepperton: Ian Allan, 1992)

Tozer, M. J., *Around the Gordano Valley* (Gloucester: Alan Sutton Publishing, 1988)

## Newspapers and Magazines

*Bath & Cheltenham Gazette, Bath Chronicle, Bristol Evening Post, Bristol Times & Mirror, Clifton Chronicle, Western Daily Press.*
*Engineering, Railway Magazine, Railway Observer.*

## Note to the Reader

Train times in this book are those given in official timetables, the twelve-hour clock being used before June 1963 and the twenty-four hour clock subsequently.

## Acknowledgements

Thanks are due to John Mann for providing information not otherwise readily available and to Colin Roberts for checking and improving the text.

# Introduction

Why should anyone wish to construct line to Portishead?

Growth is the answer. In the middle of the 19th century the size of ships grew larger making it difficult to navigate the River Avon up to Bristol, and there was always the risk of a vessel being swept sideways by the current, becoming jammed across the river as the tide fell and then breaking her back.

Another problem was that ships could only reach Bristol around the time of high tide and as this varied daily, it was impossible to run a regular shipping service while to add to the problems during winter months half the high tides occurred during the hours of darkness.

The solution was to build a port at the mouth of the Avon. Actually two such ports were constructed: one on the south side at Portishead and the other on the north bank at Avonmouth both served by railways.

The two ports flourished in the late 19th and early 20th century, becoming particularly vital to the country in both World Wars being some distance from the enemy. The post-war growth of car ownership then caused the Portishead branch to be closed to passenger traffic on 7th September, 1964 and freight declined, the last train running on 3rd April, 1981. Growth in housing and car ownership in the late 20th

Comic card drawn by Cynicus, Martin Anderson (1854-1932), of Leuchars, Fife. He was a talented, if eccentric, artist whose satirical and political cartoons did much to popularize early picture postcards. This card was produced for many small, slow, branch lines and overprinted with their name, on this example 'Bristol to Portishead'.

*Author's collection*

century and early 21st century caused serious road congestion between Portishead and Bristol during the rush hours and the obvious solution was to re-open the line to passenger traffic.

Meanwhile the opening of Royal Portbury Dock demanded rail access, so the mothballed Portishead branch was returned to working order and a link built between Pill and Portbury Dock (*see map, page 90*).

As Portishead was only just over three miles from Pill, those pressing for the line's re-opening should have had a good case, but finding the cash required proved a serious problem. At the time of writing it seems possible that passengers may be able to travel to and from Portishead by rail in 2021 or 2022.

Ashton Gate 29th June, 2001, track relaid for opening to the Royal Portbury Dock.  *Author*

# Chapter One

## Early Schemes

The first proposal for a railway to Portishead goes back to the very early days of railways when in 1800, Mr Grace, who lived at Bristol, suggested building a railway between the coal pits and his wharf at Portishead tide mill. It was to be on an inclined plane thus enabling descending loaded wagons to draw up the empties. Unfortunately this ingenious plan was not developed.

The first complete harbour scheme for Portishead was that put forward by W. C. Mylne in 1828. The project proved moribund but with the active construction of the Great Western Railway from London to Bristol, in 1839 there was a congress representing the various municipal, maritime and commercial bodies in Bristol which met to consider what should be done for the improvement of the Port when the proposal of a Pier at Portishead was adopted.

Through the years several other abortive schemes were produced and in May 1845, Isambard Kingdom Brunel, keen on innovations, in order to cope with the rise and fall of the tide which sometimes varied as much as 45 feet, planned to build a floating pier at Portbury and to connect it to Bristol by an atmospherically-worked railway.

With this system, like the contemporary one he was installing on the South Devon Railway, no locomotive was required, but a special leading vehicle would have had a piston attached to fit into a pipe fixed between the rails. When pumping stations set at intervals along the line withdrew air from this pipe, air pressure from behind would have pushed the piston and the train along.

Associated with this railway was an inclined plane worked by a stationary engine to link the proposed railway with the then unfinished Clifton Suspension Bridge.

The Portbury Pier & Railway Company obtained its Act of Parliament 9 & 10 Vict. cap. 344 of 3rd August, 1846, but was wound up in 1851 through lack of subscribers due to the depression following the Railway Mania. At a special meeting of shareholders in January 1848 James Moore, Chairman, explained that there had been a great deal of hesitation in paying calls and as works were suspended, he saw no necessity for paying. Mr Rumsey agreeing with the Chairman, could not see the line being remunerative and said steps should be taken to abandon it. This decision was wise for the atmospherically-worked South Devon Railway proved this system of traction to be a failure: the leather seal leaked; the pumping engines frequently failed; they were extravagant on the use of fuel; train braking proved a problem as the handbrake on the piston

> BRISTOL AND PORTISHEAD—BRISTOL AND SOUTH WALES UNION.    31
>
> **48.—BRISTOL AND PORTISHEAD.**
> Incorporated by 26 and 27 Vic., cap. 107 (29th June, 1863), to construct a pier at Portbury, and a railway therefrom to the Bristol and Exeter, with a branch to Portishead. Length, 10 miles. Broad gauge. Capital, 200,000*l.*, in 25*l.* shares; loans, 66,600*l.* Extra land, five acres; compulsory purchase, three years; completion of works, five years. Works in progress.
> Agreements with the Bristol and Exeter, by which that company undertakes to provide the necessary rolling stock, and to work the line in connection with their system, at 40 per cent., when they shall have reached 10,000*l.* per annum.
> CAPITAL.—This account showed an expenditure up to the 30th of June of 33,702*l.*, and the receipt of 37,600*l.*
> Meetings in February and August.
>
> *No. of Directors*—8; minimum, 5; quorum, 3 and 5.   *Qualification*, 500*l.*
>
> DIRECTORS:
> Chairman—JAMES FORD, Esq., Clarence Villa, Clifton.
> Deputy-Chairman—GEORGE HOCKE WOODWARD, Esq., Cornwallis Grove, Clifton.
>
> Richard Fry, Esq., Cotham Lawn, Bristol. | Richard Foldes, Esq., Bristol.
> Richard Hobbson, Esq., Richmond Cottage, Clifton. | Frederick Weatherly, Esq., Hillside, Portishead.
> William Watson, Esq., Weston-super-Mare. | Lewis Fry, Esq., Shannon Court, Bristol.
> OFFICERS.—Sec., J. F. R. Daniel; Engrs., McClean and Stileman, 28, Great George Street, Westminster, S.W.; Solicitors, Isaac Cooke and Sons, Bristol; Auditors, J. T. Pike and W. Tribe.
> Offices—6, Clare Street, Bristol.

Bristol & Portishead Railway's details in *Bradshaw's Railway Manual & Shareholders' Guide & Directory*, 1865.

carriage was sometimes overcome by the atmospheric pull on the piston; and as there was no means of connecting one atmospherically worked track with another, horses or locomotives were still required.

Portishead was determined not to be left off the railway map, especially as in 1849 a small stone pier had been built which was served by packet steamers. Sir J. K. McNeil, Mr Croome and others prepared plans and in 1863 the demand for low water accommodation became so pressing that engineer J. R. McClean's assistance was obtained, half-a-dozen leading merchants supported him and a company was formed. On 29th June, 1863 the Bristol & Portishead Pier & Railway Company (BPPR) obtained Parliamentary powers (26 & 27 Vict. cap. 107) to construct line from a junction with the Bristol & Exeter Railway (B&E) at Bedminster, to a pier at Portbury, a short branch running on to Portishead. The line was to be worked by the Bristol & Exeter for 60 per cent of the gross receipts, reduced to 40 per cent when receipts exceeded £10,000 a year. The Act authorized a capital of £200,000 and borrowing powers of £66,600.

It was intended to lay a branch onwards to Portishead, then only a village, but a change of plan caused the company to abandon the proposed Portbury terminus and make the proposed branch to Portishead the main line. These changes were authorized by an Act of 1866 (29 Vict. cap. 88) which enlarged the company's capital by £60,000 and its borrowing powers by £20,000.

# EARLY SCHEMES 9

The company Chairman was Alderman James Ford, also a Director of the Taff Vale Railway and the Clifton Suspension Bridge companies. Other Directors, all Bristol men, were Thomas Canning, Michael Castle, Richard Fry, Richard Fuidge, Richard Robinson, Frederick Weatherley and G. R. Woodward. The company's offices were at 6, Clare Street, Bristol and the Secretary John Francis Ranald Daniel. An interesting character, he had studied law and civil engineering, later became secretary and manager of the Bristol Athenaeum and arranged for ladies the first public examinations in the country in science, classics and languages.

Daniel was Secretary and General Manager of the Portishead docks and gas and water companies. *Circa* 1867 he moved from Tyndalls Park, Clifton, Bristol to 'Fircliff', Woodhill Road, Portishead, his house constructed by the workers who built the dock. This incredibly busy man was also an instructor in the 1st Gloucestershire Volunteer Rifle Corps and raised a company of the Gloucestershire Volunteer Artillery at Portishead, eventually reaching the rank of Honorary Major and the Volunteer Officer's decoration for long and efficient service. One of the supporters of the Weston, Clevedon & Portishead Railway, he moved his office to 70 Queen Square, Bristol, about the time he was appointed its Secretary. Later he became General Manager of the Midland & South Western Junction Railway.

The company's Engineer was John Robinson McClean. Born in Belfast 1813 and a graduate of the Royal Academical Institution, Glasgow, he entered the office of Walker & Burges and worked on improvements to the Birmingham canal. Later he was engaged as Engineer on the South Staffordshire Railway and branches. He was then appointed Chief Engineer of the Furness Railway carrying out the harbour, docks and

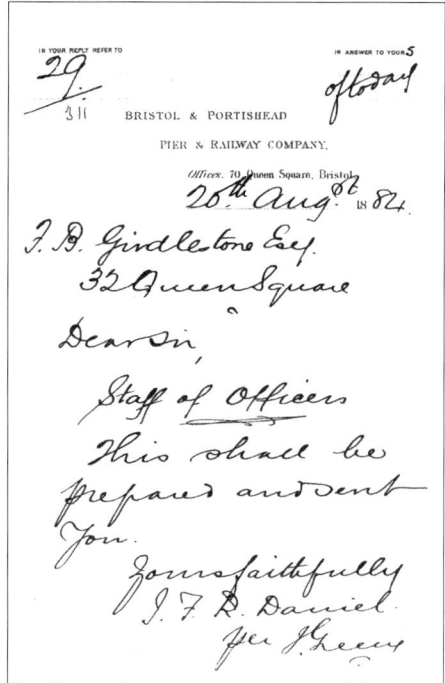

Letter from J. F. R. Daniel, Secretary, on Bristol & Portishead Pier & Railway Company's stationery.

railway improvements around Barrow. A member of the Institution of Civil Engineers, he became its President in 1865. He died at Stonehouse, Kent, 13th July, 1873 aged 60. He was assisted by Francis Croughton Stileman who had been educated at the College of Civil Engineers, Putney, articled to McClean and in 1849 was made his partner. On McClean's retirement from the Furness Railway, Stileman was appointed its Engineer. He died in London on 18th May, 1889 aged 64.

In 1869 McClean produced an abortive plans for tunnelling below the Avon at Rownham Ferry to Hotwells in order to link with the Bristol Port & Pier Railway, (an isolated line from Clifton (later Hotwells) to Avonmouth) on the other bank of the Avon.

The Portishead scheme was a rival to the Bristol Port Railway & Pier which had been authorized the previous year and opened in 1865.

Not everyone welcomed the Bristol & Portishead line. Conservationists were concerned that in the Avon Gorge it would damage the beauty of Leigh Woods, while another criticism was that opposite Hotwells three chocolate houses would have to be demolished to make room for the railway. These refreshment rooms were highly popular with the working classes during the summer and their destruction demanded substantial compensation.

Clifton Suspension Bridge which was being completed as a memorial to Brunel, was protected by a clause in the Act ensuring that the 59-yard long tunnel under one of the bridge's abutments, was built to the satisfaction of the bridge company's engineer. A benefit of the branch was revealed in the *Bristol Times & Mirror* of 16th September, 1863 when it observed that Clifton residents 'Desiring to proceed at once to London, Exeter, or the North, will only have to cross the bridge with their portmanteaux, descend some steps to the platform of the railway and walk into the carriage as it stops on its way from Portishead to Temple Meads thus avoiding having to pass through busy Bristol streets'.

In due course a new public road linked Clifton Bridge station with the suspension bridge paid for jointly by the Leigh Woods Land Company, the bridge company and the Portishead Railway. A horse omnibus ran from Tyndall's Park Road through Clifton, over the bridge and down to Clifton Bridge station.

At the first half-yearly meeting of the BPPR held on 27th February, 1864 the report was given:

Gentlemen,
The line has been resurveyed and set out since the passing of the Act, and property plans prepared for the land throughout.

The contract drawings and specifications are completed, and works are ready to be let to a contractor as soon as sufficient land is obtained to enable him to proceed with the works.

Some delay occurred in consequence of a deviation having been surveyed through Leigh Court, which would have been beneficial to the public; but as it was not agreed to, the Line within the Parliamentary limits had to be set out again, and new property plans prepared.

We are, Gentlemen, your obedient Servants,
McCLEAN & STILEMAN

William Tredwell of Handsworth won the contract for building the line, his offer being accepted on 27th April, 1864. Work started in the spring of 1864 and unusually there was no ceremony of cutting the first sod, an event which Victorians usually enjoyed. Tredwell's engineer was Mr Godsell and his resident engineer Mr Macpherson, the railway company's resident engineers being Mr Drysdale and Mr Curry. By June 1864 subscriptions totalled £37,600 while expenditure was just over £33,700.

On 22nd June, 1864 the *Clifton Chronicle* reported on the line's progress. Men were at work from Rownham to the river's mouth making preparations for fencing the line, while work on an embankment and cutting near the Rownham Inn was proceeding. The foundations of a bridge at the entrance to Nightingale Valley had been made and 500 yards below that the facing of No. 3 tunnel was being formed. Work on facing at the Pill end of No. 2 tunnel had commenced and the centre shaft of No. 1 tunnel was being sunk. Great care was taken of the environment and very few trees in the neighbourhood of Leigh Woods were cut down. Unusually the contractor was instructed by the Directors 'Not to cut down a single tree more than is absolutely necessary and to endeavour, where they find it practicable, to take the permanent way round the inner side of any large trees which might come close to the line, the foliage of these trees will then be left undisturbed, and will screen the line from the roadway on the margin of the river'.

A week later the *Clifton Chronicle* disclosed:

> The contractor for the construction of the Portishead line seems determined to carry out the works with vigour. Although the line was only recently commenced, large numbers of men are now engaged in making six cuttings and two tunnels between Rownham and Pill and fixing fences, and erecting workshops. In addition to this, men have been engaged night and day during the past week in forming a small dock a little below Rownham to facilitate the landing of material required on the work.

At the company's half-yearly meeting on 31st August, 1864 it was announced that works had commenced and were 'in course of vigorous prosecution'. Most of the land had been purchased except for a few acres not required for immediate use. A fair proportion of the purchase money was in the form of a rent charge by the principal landowners. Expenses to June had been £33,702 and receipts £37,000.

Tredwell was starting on the works which would consume the greatest amount of time: tunnels, deep rock cuttings, bridges and heavy cuttings and embankments. The Directors hoped to open the length from the junction with the Bristol & Exeter to Clifton Bridge early in 1865, the B&E agreeing to work this initial length. Some people had criticized the railway for spoiling Leigh Woods whereas the Directors claimed that the presence of the line would stop quarrying and blasting there and thus preserve, rather than spoil, the amenities.

The amenities of the Ham Green Estate were to be preserved by carrying the railway in two tunnels, but in the event, the tunnel east of Ham Green Halt was replaced by a cutting. Altogether there were four tunnels on the branch making a total length of 1,044 yards.

At the half-yearly meeting on 28th February, 1865 it was announced that despite severe weather, works had advanced more than anticipated and from Portishead Junction (later Parson Street Jn) to Clifton

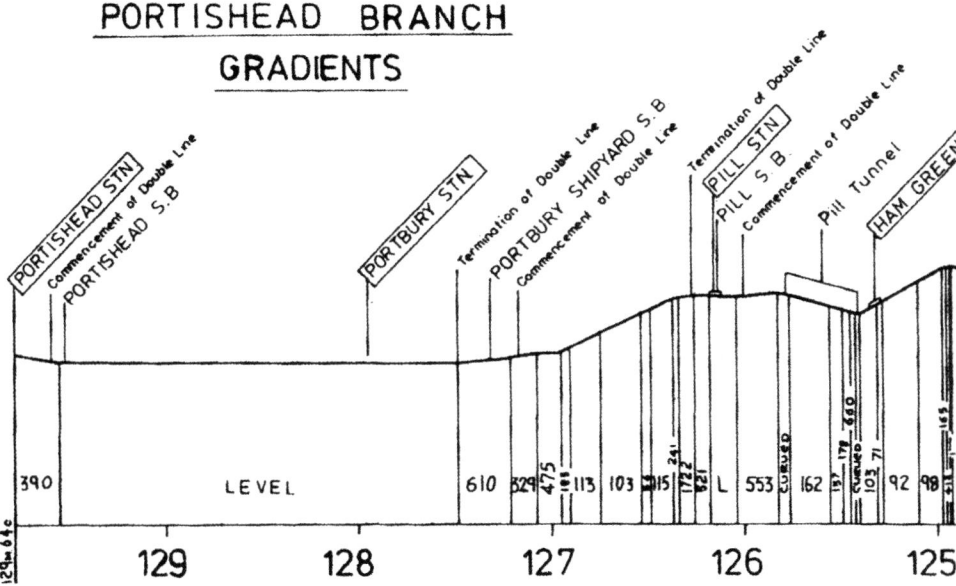

# EARLY SCHEMES

Suspension Bridge the formation was ready to receive ballast and permanent way. From the bridge to west of Pill much heavy work had been completed. Sir William Miles had allowed an open cutting to replace a long tunnel through Leigh Woods thus saving cash.

In April 1865 it was believed that the line would be completed within a year. Messrs Crawshay, Bailey & Co. were to supply the rails, Mr Weston of Redcliff Street, Bristol the fishplates and bolts, while Mr Batchelor, of Newport was to supply the timber. [Redcliffe Way and the church of St Mary's Redcliffe are so spelt, but other roads omit the final 'e'].

The *Bristol Times & Mirror* for 13th April, 1865 aroused enthusiasm:

> For the first two miles of the journey from the junction of the Bristol & Exeter Railway to Rownham [Clifton Bridge], the passenger will travel through delightful scenery. From Rownham the line passes for two and a half miles between the beautiful woods of Leigh on one hand and the river on the other, while it passes through Pill and continues its course through verdant pastures until it reaches Portishead. The company are indebted to four baronets and to other people for the ready surrender of the necessary lands.

The rival *Western Daily Press* for 13th April, 1865 was less enthusiastic saying that the old pleasure gardens at Rownham 'have lost their beauty, the trees hang down their heads as if mourning their loss, and

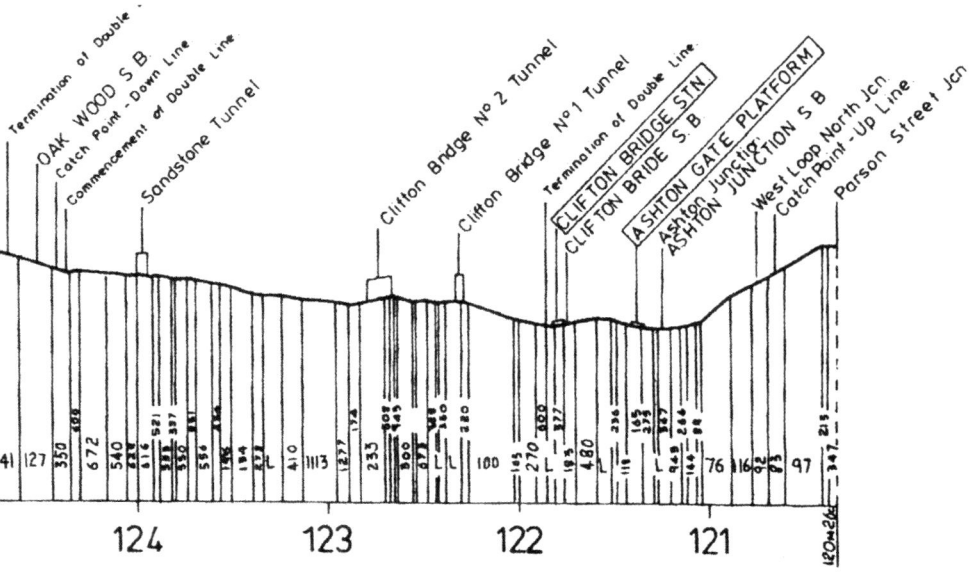

the bowers and arbours are in a state of sickening dilapidation'. What had been the New Inn & Posting House had been converted into the engineers' office occupied by the resident engineers Messrs Drysdale & Curry and the contractor's engineer Mr Macpherson. The contractor Mr Tredwell, through his representative Mr Godsell was 'pushing forward the work with great assiduity'. The men were chiefly engaged on the Portishead to Rownham section trimming the slopes to be in a condition to be overlaid with grass turf. Ashton Road near the Ashton Vale works, had been diverted for about 250 yards.

In April 1865 other navvies were busy blasting Clifton No. 1 Tunnel through the rock below the suspension bridge, the latter having been opened to the public on 9th December, 1864. Three holes were bored and charged with powder; the fuse ignited and the alarm given, the men scrambling to get outside; then one explosion was followed by two more.

A report revealed that beyond the junction with the Bristol & Exeter was the steepest gradient – 1 in 97 down. Although only single track from the junction to Clifton Bridge, the overbridge allowed for double.

The railway passed through the middle of the Ashton Vale Iron & Coal Works, an arrangement beneficial to both companies as coal from the mine could be sent direct by rail instead of being carted along Ashton Road to the railway. The Iron & Coal company constructed its own private lines on both side of the branch.

Four hundred navvies moved to Pill for making the tunnels, cuttings and the viaduct. Very deep excavations were required for the foundations of Pill viaduct and due to shifting sand were strengthened with bags of cement. Some of the navvies sought accommodation in cottages, the villagers appreciating the rental obtained, while other navvies occupied huts on the site of Monmouth Road by the station.

Approaching Portishead was a long, curved timber viaduct with 23 spans supported on trestles 25 feet apart. Its underside was about five feet above the water-line

The navvies being strong, muscular and relatively well-paid, were popular with the girls of Pill parish and this amorousness led to a fight between the local lads and navvies on Whit Monday 1863. It started at 2.00 pm and PC Fry stopped it at 4.30 pm.

At the half-yearly meeting in August 1865 it was reported that works were steadily progressing and it was expected that the whole line would be ready for the permanent way by February 1866. Calls on the shares had been paid punctually – a claim many companies could not make but would have enjoyed doing so.

# Chapter Two

# The Line Opens

When the railway contract was completed, most of the navvies moved on, though a few remained. Although the contractor had to contend with the weather – two long, severe winters and an unusually wet summer, he had not suspended works.

The line finished, it was inspected on 12th April, 1867 by Colonel Yolland on behalf of the Board of Trade to ensure that it was safe to be opened to public passenger traffic. The single track, broad gauge line 9 miles 11½ chains in length had stations at Clifton Bridge, Pill, Portbury and Portishead and a siding at Ashton Vale Works. Curiously all the stations were in alphabetical order. At the far end of the line was a 48½ chain extension on to the pier at Portishead. The flat-bottomed rails were fitted directly to the sleepers with screw bolts and fang nuts. A turntable was provided at Portishead, while at the other end of the line, the Bristol & Exeter's turntable at Temple Meads could be used. The branch had 11 over and 10 underbridges, three viaducts and four tunnels. Most of the bridges and viaducts were of stone and brick. Colonel Yolland was concerned that some of the fencing permitted livestock to stray on the track and ruled that more intermediate fence posts be inserted. He said that signalling at the junction with the Bristol & Exeter and at the Ashton Vale siding was inadequate. The line was too close to two of the overbridges and the retaining walls were required to be cut back. In No. 2 tunnel an open carriage door fouled the tunnel wall and an underbridge between Portbury and Portishead needed strengthening. At Portbury he required the facing down siding to join the up siding rather than the down road in order to avoid a facing point and wanted a platform for up trains. These alterations were carried out, the company informing him of this on 16th April, 1867. Despite these minor criticisms Yolland said he was 'well satisfied with the effective and substantial way in which the works had been designed and executed'.

He described the railway:

> The line is single throughout with loop sidings, or passing places, at the stations at Clifton, Pill, Portbury and Portishead, but all land has been purchased and the overbridges and one of the tunnels have been constructed for double line if hereafter required. The gauge is 7 ft 0¼ in. and the width between the lines is 6 ft.
>
> There are 11 over and 10 under bridges, three viaducts and four tunnels. The over and under bridges are all constructed with stone and brick, with the exception of 3 or 4 which have cast iron platforms to carry the roadway. They are well-constructed and sufficiently strong, the largest span is 40 ft on the skew.

A Portishead to Temple Meads train leaves No. 1 tunnel 4th June, 1960. Notice the fixed distant signal, *right*. *R. E. Toop*

The tunnels are respectively 66, 231, 88 and 660 yds in length. [Modern measuring has slightly modified these lengths.] The first is for a double line, lined throughout with brick and stone, the second through limestone rock and only partially lined, the third is through sandstone and lined with brickwork in mortar. The fourth is through marl and is lined throughout with brickwork in mortar. There is one ventilation shaft for this tunnel and some water passes through the brickwork. The whole is well constructed.

The Clifton Bridge Company's chief engineer inspected the tunnel below his bridge's abutment and found it satisfactory.

The 9 miles 49 chains of broad gauge line from the junction with the Bristol & Exeter Railway, opened on Thursday 18th April, 1867 and even before the line was inaugurated, 'numerous applications had been made for season tickets'. Unusually for Victorian times, there were no special opening celebrations. Goods traffic was carried by passenger trains until about July 1875. The total cost of the line was £185,000 comprising £117,000 for works, £47,000 for land and £21,000 legal and other expenses.

Regarding the lack of celebrations the *Bristol Times & Mirror* for 20th April, 1867 commented:

> Another somewhat singular circumstance in connection with the concern is that it opened without a banquet; that not a champagne cork has been popped

## THE LINE OPENS

off or a cannon fired, a shilling expended for wine or for powder (save so far as the latter was required for blasting the Leigh rocks) from the turning of the first sod to the starting of the first train. There is an Irish adage which says 'a dry christening forbodes a dull baby' and we can only hope the hospitable Bristolians will forgive the omission in this case, and that it will be no ill omen to the undertaking that its inauguration has not be renowned by even an ounce of cheese or a pint of beer.

At the company's half-yearly meeting on 29th August, 1867 the Directors said that the Bristol & Exeter was working the line 'in the most economical manner consistent with complete efficiency' charging only for the cost of working. By the articles of agreement, the Bristol & Exeter had not been obliged to work the line until the pier opened, but had kindly consented to work the line immediately the railway was completed.

The timetable, which the Bristol & Exeter's traffic manager had assisted in making, was suitable for the convenience of temporary and permanent residents of Portishead as well as for daily visitors and excursionists. Between the opening on 18th April and 30th June, over 40,000 passengers had been carried and receipts averaged almost £30 per working day. Traffic to and from the pier was regarded as the principal source of revenue and pier works in progress would be 'pushed on'. The engineer McClean said that he had not known during the whole of his professional career a line taking such large mileage receipts shortly after the line had been first opened.

At the half-yearly meeting on 29th August, 1867 some shareholders criticized the fact that not sufficient amusements, like the pleasure garden as at Avonmouth, had been offered to induce the public to use the line. The Directors replied by saying that they felt that the line's natural beauty was sufficient and hoped that, as at other watering-places, building houses would make Portishead would grow like Clevedon and Weston-super-Mare.

At the half-yearly meeting held on 28th February, 1868 it was announced that during the previous half-year 119,439 passengers had been carried: 10,703 first class, 102,092 second and 6,644 third; these figures excluded 48 season tickets. (The fact that third class tickets were the least popular is interesting.) Receipts totalled £4,509, working expenses being just over £3,000. By 1st May, 1868 the Directors expected to be able to deal with passengers, cattle and merchandise at fixed hours daily, irrespective of the state of the tide. McClean and Stileman proposed that for a relatively small outlay, a 30 acre dock should be made.

The railway operated steamer services to Cardiff and Newport – the railway route to South Wales hitherto being via Gloucester as the Severn Tunnel was not opened until 1886. In summer ships plied to Ilfracombe with through rail and steamer bookings from the Great Western and the Midland railways.

At the half-yearly meeting held on 26th August, 1868 it was announced that in the half year 68,965 passengers had been carried, excluding 64 season ticket holders. A year later, despite the fact that wet and cold weather had a detrimental effect on holiday traffic in the first half of 1869, the figure had increased to 75,701 while the half-year's revenue had increased by almost 40 per cent.

Some 94,069 passengers, excluding season ticket holders, were carried in the half-year ending 30th June, 1872. The gross revenue of £4,099 8s. 11d. showed a small increase over the previous year but no dividend was declared. At the half-yearly meeting on 28th August, 1873 figures for the first six months had been 105,560 bringing in a revenue of £4,259. Enhanced prices of material and labour had made maintenance more costly, but careful economy in other departments enabled the company to show a reduction in working expenses. As early as August 1873 the Directors realized that the gauge would need to be narrowed.

Bristol & Exeter Railway Pill to Portishead ticket issued 30th May, 1871.     *M.J. Tozer*

# Chapter Three

## Portishead Pier & Shipping

The first 200 feet of the pier was constructed of concrete blocks, with timber for the remainder. The wooden portion started almost at right angles to the concrete portion and projected into the Channel, with one side to a considerable extent, land-sheltered. The concrete blocks measuring 8 feet by 4 feet by 3 feet each weighed 8 tons. Made on site, they were composed of local sand, Portland cement, broken stone and water. The concrete was similar to that used for the Admiralty Pier at Dover.

The ingredients were first mixed on a small platform and then shovelled into a wooden mould-box, where they were allowed to remain for two days by which time they had become sufficiently hard to retain their shape. The sides of the mould were then knocked away and the blocks exposed to the atmosphere and became hardened in about a month. They were placed on trucks and taken to the pier staging where a travelling crane lowered them into position. The upper deck of the pier had two railway tracks also used by a 5-ton steam crane. The pier was lit by gas. Initially wagons and carriages reached the pier by turntable, but later an embankment was made to enable them to run directly on to the pier.

The solid part of the pier resting on a rock foundation, started at a height of 10 feet from the beach at the shore end and at its extremity rose 35 feet above the foundation. The footway on the seaward side of the pier was chiefly formed by the upper surface of the concrete blocks, while the footway on the inner side was of wood. The pine deck planking was 4 inches thick except where the rails ran where a bedding was 4 inches of English oak. The pier was 30 feet wide and its balustrade

Portishead Pier 1870                                      *Author's collection*

Portishead Pier circa 1905, from the south looking at the western side of the pier.
*Author's collection*

The upper deck of Portishead Pier circa 1905. Looking north over the eastern side of the pier. The piles heads with their chains fence the edge. *Author's collection*

formed from the pile heads which protruded 3 feet above deck level. On the western side were massive wooden hand-rails which could be removed to allow gangways to be placed, while on the eastern side was a double row of chains which also could be removed without difficulty.

The bulk of the steamboat traffic was worked from the eastern side of the jetty where there were as many as 24 separate passenger landings at different levels and at various points in the length of the pier below the upper deck. There were also 16 cattle landings similarly distributed. The passenger landings communicated with each other and with the deck by convenient timber stairways. Communication between the cattle landings was by ramps with cross-battens to prevent slipping, gates being placed to guide beasts the correct way. There were also eight landings on the western side available for passengers or cattle.

On the upper deck of the pier were two lines of rails for the purpose of transferring goods from the steamer to the railway or vice versa. A rail-mounted 5-ton steam crane is provided.

At the half-yearly meeting on 28th February, 1868 it was announced that by 1st May works on the pier would be sufficiently advanced for steamers to land and embark passengers, goods and cattle to and from the railway at fixed hours, irrespective of the tide and without interference to building work. This estimate proved slightly ambitious and it was not until June 1868 that the timber pier, as opposed to the jetty, was opened at Portishead, 540 ft in length and carrying double track.

Before the half-yearly meeting on 26th August, 1868, the Directors and shareholders inspected their new pier where from the opening, for 18 hours out of 24, passengers, cattle and merchandise could be landed. The cattle landing stage was below that used for passengers and so could be driven screened from view until they reached commodious pens where they were held until loaded on a train. Water was provided in troughs and globular gas lights were provided down the centre of the top deck and suspended over the gangways and stairways below; so that passengers and cattle could be landed at night. The heads of the stairways could be closed by hatches when wagons or carriages were required to run over them. While the shareholders were inspecting their new pier, a steamer came alongside, loaded passengers and was away in three minutes.

The company had carried 188,404 passengers during the half-year, excluding 64 season ticket holders and over the previous 10 weeks 4,000 passengers had used the pier, 'in many instances effecting a through journey in from three to eight hours' less time than it would otherwise have occupied had they been required to wait for the tide. For instance, on 22nd August, 1868, the *Apollo* from Cork landed passengers at Portishead who reached London that same night, whereas but for Portishead pier they would have had to wait for the tide and the next day being Sunday, they could not have reached London until evening. The passage from Cork had been made in 22-23 hours instead of the usual 30 hours.

In April 1869 the Cardiff & Portishead Steamship Company Limited of Cardiff ran services from both Bristol and Portishead piers, but that July the Newport run was suspended and three paddle steamers, *Taff, Dart* and *Ely* placed on the daily Cardiff run, but then that service was abandoned in October.

At the meeting on 27th August, 1869 it was announced that Tredwell, the contractor for building the railway and the first section of the pier, had been selected to build the 300 ft pier extension. Frank G. Stileman was engineer to the pier project. Traffic on the railway continued to increase.

Regarding the experimental service between the main line at Temple Meads and steamers run at a fixed time irrespective of tides, 'although attended by many difficulties and inconveniences arising from the incomplete state of the works, [they] had succeeded in developing a passenger traffic exceeding the most sanguine expectations'. The cattle landing stage had proved extremely efficient and a vessel from Cork had landed 500 head of cattle and was away again in 25 minutes.

Before the meeting on 27th February, 1870 shareholders inspected the pier works finding them 'in a very forward state' and almost complete. Despite very stormy weather the works had not been affected 'not a stone had been started, or a bolt moved'. When finished double track will be laid along it to enable passenger carriages and wagons to draw alongside steamers instead of terminating by the refreshment room. The pier would be able to accommodate simultaneously three of the largest Irish steamers and would enable vessels to discharge on the west side as well as the east.

During the half-year 114,614 railway passengers had been carried, excluding season ticket holders, bringing in a revenue of £4,562 10s. 5d. against 82,551 and £3,259 13s. 4d. in the corresponding half of 1868, an increase of 40 per cent. Double the number of season tickets had been issued.

Train miles for the half-year were 22,724 with working costs of £3,696 10s. 2d. against 21,590 and £3,124 5s. 9d. of 1868. Most of the increased expenditure was due to opening the new route to Ilfracombe and Lynmouth via Portishead. Arrangements had been made with the GWR 'by which a large number of passengers were conveyed to and from Ilfracombe by train and steamer' and the Directors believed that these numbers would increase further. Michael Castle, Deputy Chairman of the Bristol & Exeter, observed that if third rail was laid, the Portishead line could connect with the Midland Railway.

On 18th April, 1870 a 300 ft-long low water extension was added to the pier, extending it into deep water to give ships access at any state of the tide. A double-track railway ran along its original length of 540 ft. The company's policy of carrying out work in stages: first the railway, then a short pier, and then a deep water pier, had proved successful.

The *Illustrated Midland News* of 17th September, 1870 reported:

> Steam vessels ply every day at certain hours, independent of tide, to Ilfracombe and Cardiff, and at short fixed intervals to Cork, Dublin, Waterford and Tenby, starting on the arrival of the trains at the pier and bringing through passengers from the Great Western Railway, the Midland, and the Bristol and Exeter. Since the opening of the Pier there has been an increase of something like five-hundred per cent, upon the steam-boat traffic to and from the North Devon and Welsh ports via Bristol This route to Ireland has special advantages, for passengers from Paddington can make the whole run in twenty hours for a first class fare of about 40s., while from Birmingham and the Midlands the run is shorter and cheaper still, and in the journey there is only one change-which is a very easy one – from the railway carriage at the end of the pier to the steam vessel lying alongside.
>
> To the traveller taking this route, the change from the old mode of getting away from Bristol to the new is very striking. Until now he had to scramble for his luggage at Bristol station, and jolt three miles through Bristol in a cab, counting himself fortunate if he did not miss the steamer; and then there was the paddle at the comparatively slow rate down the inconvenient seven miles of river. Under the most favourable circumstances the journey occupied half as much time again as it does now, and was considerably more expensive.

In September 1870 excavations for widening and deepening the Pill were in progress. A powerful steam dredger, with a set of hopper barges cleared mud away at the rate of 1,000 cubic yards a day to create a minimum depth of 6 feet with a maximum width of 50 feet alongside the pier, thus making the pier independent of tides and allowing channel steamers to come alongside at any state of the tide.

At a special meeting of the company held on 14th March, 1871 it was announced that the use of the pier by cargo steamers and sailing vessels in addition to passenger ships was increasing.

In 1873 the *Ely* and *Taff* were purchased by the Portishead Steamship Company. The Lyn Steamship Company was an offshoot of the

Portishead Steamship Company which in turn was closely related to the Portishead Dock Company of which J. F. R. Daniel was the managing owner.

On the morning of 15th February, 1874 the temporary dam holding back the sea water collapsed, meaning that work from then onwards was dependent upon the tide, but by the end of the year it was anticipated that the entrance lock would be finished so the tide could be excluded from the inner portion of the works and the contractor no longer constricted by the tides.

Inland from the pier and in line with it, a dock was constructed with an entrance dock. For ships waiting to enter a convenient anchorage ground was at King Road, adjacent to the entrance. *Lyn* was the first vessel to enter Portishead Dock, this event, with no opening ceremony, taking place on 28th June, 1879, *Lyn* continued to work the service until 1885. Each summer *Ely* and *Taff* were utilized on the Portishead-Ilfracombe run. The *Western Daily Press* revealed that on Bank Holiday Monday, 4th August, 1873 *Ely* was on the down channel service and that Portishead Pier was open. On some days services were run to Cork, Dublin, Waterford and Tenby starting on the arrival of trains on the pier bringing passengers from the GWR, MR and Bristol & Exeter. A year following the opening of the pier, traffic to and from North Devon and the Welsh ports had increased by 500 per cent.

This route to Ireland had the advantage that passengers from Paddington could make the whole journey in 20 hours for a first class fare of about £2. A passenger simply stepped from the railway carriage at the end of the pier into the steamer, instead of having to arrange transport from Temple Meads and pass through the streets to Bristol Harbour with the possibility of not reaching the vessel before it sailed. The ship then had to navigate the tortuous seven miles of river to the Bristol Channel at a relatively low speed. The railway and pier enjoyed 250,000 passengers annually producing a gross income of £10,000.

*Taff* was sold in 1878 and *Ely* in 1880, the company then ceasing to exist as owners of packet ships, but still possessed a screw-driven passenger tender *Bee*, 31 tons gross. She had been built for J. F. R. Daniel in 1870, purchased from him in 1877 and eventually sold to Bristol Corporation in 1912. In 1885-1886 *Gael* 347 tons gross, owned by the GWR worked from Portishead to Ilfracombe before the GWR abandoned the service on 1st October, 1886. Rails on the pier were lifted some time prior to 1928. About 80 feet of the pier was damaged on 10th February, 1956 by the SS *Astro*.

# PORTISHEAD PIER & SHIPPING 25

Portishead Pier 29th March, 1956. *BR*

Portishead Pier 29th March, 1956. *BR*

From 1877 coal for Portishead gas works was unloaded at the pier into railway wagons and drawn to the works. The signals controlling the siding, the siding itself and the connection were maintained at the gas company's expense. The gasworks manager was granted a free pass between Portishead and Bristol and the company gave the railway a 10 per cent discount on the gas it used. In 1931 these terms were discontinued and the works closed shortly after the Second World War.

The site of Portishead Pier station is now occupied by the lifeboat station, built using some of the original materials.

Portishead Pier station as a masonic temple *circa* 1980. *Author's collection*

# Chapter Four

# Portishead Dock

Portishead harbour had been used from time immemorial as a place of shelter from heavy gales, its sheltered position rendering it the safest place in the Bristol Channel. At the half-yearly meeting on 28th February, 1868 the Bristol & Portishead Pier & Railway Company's engineers McClean and Stileman reported that for a comparatively small outlay, 30 acres of water by Portishead Pill could become tidal harbour or floating dock. (A 'pill' is the local term for a promontory.)

The idea was adopted and in 1871 a Bill promoted (34 & 35 Vict. cap. 142) to make Portishead Pill into a dock covering 12½ acres to a depth of 24 ft. As the Avonmouth Company on the opposite bank of the river also had plans that year, both companies were competing for a subscription of £100,000 from Bristol Corporation. The Portishead Act 34 & 35 Vict. cap. 142 allowed Bristol Corporation to purchase the dock at cost price before 1874 on condition that it subscribed the £100,000.

The contract for building the dock was let to Barnett & Gale at £160,000 on 20th July, 1872, construction starting in January 1873. Stone for constructing the dock wall was brought from Portishead by a

Portishead Dock from the air in the 1920s. Portishead station is on the left of the photograph, the railway crosses the southern end of the dock on an embankment and continues past the gas works (*centre*). The timber yards are on the right-hand side of the dock.

*Oakwood collection*

tramway. On Sunday 15th February, 1874 a high tide caused the centre portion of the coffer dam to collapse. This meant that progress was now restricted by the tide level. It was fortunate for the contractor that it happened before work was commenced on the dock basin as it allowed him to take extra precautionary measures for the safety and progress of the works. The lock wall 400 feet in length, a splendid piece of masonry had been completed to its full height and coped with granite blocks two to four tons in weight. At the half-yearly meeting on 29th August, 1874 it was announced that it was hoped to finish the entrance lock by December which would allow the tide to be excluded from the inner portion of the work and enable the contractor to work without hindrance. The mortar used was lime made from Aberthaw pebbles. The contractor used locomotives to assist work. A pleasing feature was that all the calls except one had been paid, whereas many railway companies found that quite a few shareholders defaulted.

Stormy weather in November and December 1874 slowed progress on the dock. At the half-yearly meeting 27th February, 1875 the engineer F. C. Stileman reported that considerable progress had been made on the dock.

On 10th June, 1875 the lock entrance was temporarily closed by iron girders, planked on the outside. The contractor was now able to lay rails on both sides of the pill into the dock and found that the alluvial deposit needed time to harden in order to support the locomotives and wagons. He was then able to excavate down to what would become the bottom of the dock. At the meeting on 29th February, 1876 it was revealed that 60,000 cubic yards had been excavated from the dock and although wet weather had delayed work, the lock was within 3 feet of foundation level. The lock gates were being fabricated by Messrs Maudsley, Cardiff. It was anticipated that the docks would open in the summer.

News was much better on 30th August, 1876 as 'very considerable progress' had been made with the dock works and but for the accident on 15th February, 1874 the docks would have been finished that summer. A shortage of manpower was being experienced. On 28th February, 1877 shareholders were informed that the lock gates were being delivered.

On 28th August, 1877 shareholders received the unwelcome news that dock works had suffered through the long and very wet spring having seriously delayed works. 45,000 cubic yards were being excavated each month. The average number of men employed formerly 580, had increased to 800.

An Act for additional powers received Royal Assent on 12th July, 1877. In March 1878 excitement was growing and it was anticipated that

# ILFRACOMBE,

## Via BRISTOL AND PORTISHEAD, AND THENCE BY STEAMER.

Every Tuesday, Thursday, and Saturday, from May 23rd until June 30th, and every day (except Sundays) after that date, the Portishead Steamship Company's First Class Paddle Steamers will leave Portishead for Ilfracombe direct, at about 1.30 p.m., and Ilfracombe for Portishead every Monday, Wednesday, and Friday, from May 23rd until June 30th, and every day (except Sundays) after that date, in time for the Up Trains, and First, Second, and Third Class Tourist Tickets available for One Calendar Month, will be issued via Bristol and Portishead in connection with the Boats from the undermentioned Stations at the Low Fares named :—

| STATIONS | To and from Ilfracombe, by Railway & Steamer to and from Portishead. | | Circular Tour—By Rail to Bir'st. pie. and thence by a w'll appointed Service of Car'.es to Ilfracombe and Back to Bristol by Steamer via Portishead, or vice versa | | STATIONS | To and from Ilfracombe, by Railway & Steamer to and from Portishead. | | Circular Tour—By Rail to B'rm.ngh,and thence by a well appointed Service of Carr ges to Ilfracombe and back to Bristol by Steamer via Portishead, or vice versa |
|---|---|---|---|---|---|---|---|---|
| | Tourist Tickets, available One Ca'endar Month. | | | | | Tourist Tickets available One Calendar Month. | | |
| | 1 cl. | 2 cl. | 1 cl. | 2 cl. | | 1 cl. | 2 cl. | 1 cl. | 2 cl. |
| FROM | s. d. | s. d. | s. d. | s. d. | FROM | s. d. | s. d. | s. d. | s. d. |
| BATH | 16 0 | 14 0 | 25 6 | 21 6 | Leeds | 73 6 | 56 6 | 85 6 | 66 0 |
| STONEHOUSE | 27 6 | 19 0 | 31 6 | 26 0 | Bradford, SHIPLEY | 73 6 | 56 6 | 85 6 | 66 0 |
| GLOUCESTER | 25 6 | 21 6 | 15 6 | 28 6 | Nottingham | 59 6 | 47 0 | 31 6 | 53 6 |
| CHELTENHAM | 28 0 | 16 0 | 36 6 | 30 0 | Leicester | 53 6 | 42 0 | 27 6 | 49 0 |
| GREAT MALVERN | 33 6 | 23 6 | 30 6 | 43 6 | York | 85 0 | 65 0 | 42 0 | 91 6 | 71 6 |
| WORCESTER | 31 6 | 28 0 | 19 6 | 43 0 | 35 0 | Hull | 86 0 | 67 0 | 41 6 | 93 6 | 74 6 |
| *Birmingham | 43 0 | 34 6 | 22 0 | 52 6 | 41 6 | Cambridge | 60 0 | 47 6 | 33 0 | 68 6 | 54 6 |
| TAMWORTH | 48 0 | 38 0 | 26 0 | 56 6 | 45 0 | Northampton | 58 6 | 46 6 | 30 6 | 70 0 | 53 6 |
| BURTON | 51 6 | 41 0 | 17 0 | 60 0 | 48 0 | PETERBORO' | 65 0 | 50 6 | 33 0 | 74 0 | 57 0 |
| Derby | 56 0 | 44 0 | 29 0 | 64 6 | 52 0 | *L. & Y. Stations, via Oakenshaw. | | | | |
| *Liverpool | 61 6 | 49 0 | 29 0 | 76 0 | 60 0 | | | | | |
| *Manchester Via Matlock | 65 0 | 49 6 | 31 0 | 76 0 | 57 0 | Halifax | 73 0 | 56 6 | 15 0 | ... | ... |
| Stockport | 62 6 | 48 0 | 19 0 | 75 0 | 57 0 | Huddersfield | 70 0 | 53 6 | 33 6 | ... | ... |
| Chesterfield | 54 0 | 50 6 | 31 0 | 72 6 | 57 0 | Rochdale | 71 0 | 57 6 | 34 6 | ... | ... |
| Sheffield | 67 6 | 56 53 0 | 34 6 | 76 0 | 60 0 | Bolton | 66 6 | 52 0 | 32 0 | ... | ... |
| Masboro' | 67 6 | 53 0 | 31 0 | 74 6 | 57 0 | Wigan | 67 0 | 52 0 | 30 6 | ... | ... |
| Rotherham | 67 6 | 53 0 | 34 6 | 77 0 | 60 6 | Blackburn | 70 0 | 55 0 | 33 0 | ... | ... |
| Barnsley | 71 6 | 56 6 | 37 0 | 80 6 | 63 6 | Chorley | 74 6 | 16 0 | 36 6 | ... | ... |
| Wakefield | 73 6 | 56 6 | 35 9 | 83 0 | 65 0 | Southport | 76 6 | 57 6 | 37 6 | ... | ... |

* In Liverpool Tickets are issued at Cook's Excursion Office, 14. Cases Street (opposite Central Station), and at the New Central Station. In Manchester at Cook's Tourist Office, 45, Piccadilly, and at the Midland Booking Office, London Road Station. In Birmingham at Cook's Excursion Office, 16, Stephenson Place, and at the Midland Booking Office, New Street Station.

Children under Twelve Half-fare.

THE STEAMERS, VIA PORTISHEAD, SAIL IN CONNECTION WITH THE TRAIN LEAVING

Derby at 6.55 a.m. Birmingham 9 5 a.m. Worcester 10.12 a.m. Cheltenham 10 47 a.m. Gloucester 11.17 a.m.

thus enabling Passengers from these places to get through in one day.

Passengers may land at or return from Lynmouth, at which place the Steamers to and from Ilfracombe call regularly.

At Ilfracombe and Lynmouth there are First-Class Hotels and Excellent Boarding Houses.

The trains to Portishead leave the same Station at Bristol that Passengers arrive at, and Luggage is transferred to and from the Trains, and also to and from the Steamers at Portishead, free of charge. The Fares include Landing Fees and Tolls at Ilfracombe and Lynmouth, and Steward's Fees, and those for the Circular Tour also include fees to Coachmen, Guards, &c.

Passengers from Stations North of Derby or South of Leicester, may break the Journey at Birmingham or Bristol, also from Stations at Bristol to Portishead for a period not exceeding three days.

TICKETS FOR CIRCULAR TOUR.—These Tickets will not be issued until June 1st, when the Carriages commence running between Barnstaple and Ilfracombe, for particulars of which see page 17.

The Railway Journey from Bristol to Portishead affords fine views of the Avon, Clifton Suspension Bridge, Leigh Woods, St. Vincent's Rocks, Portbury, and Portishead Woods and Harbour. The voyage in first-c ass Steamers is through the most picturesque part of the Bristol Channel, by Woodhill and Walton Bays, Clevedon, Weston, Brean Down, the Holmes, Minehead, Bossington Hills, Porlock Bay, Gore Hill, the Foreland, Glenthorne to Lynmouth, thence passing Lee Bay, High Veer, Heddon's Mouth, and Combe Martin Bay, to Ilfracombe. Mortehoe, Morte Bay, Woolacombe Sands, Barricane, the Tors, Watermouth Caverns, &c., are within walking distance of Ilfracombe; while from Lynmouth, Lynton, Lynclife, the Valley of the Rocks, the Valleys of Lynns, Waters-Meet, Glenthorne, and the Wilds of Exmoor, are easily accessible.

# TOUR THROUGH THE SNOWDON · DISTRICT.

## FIRST AND SECOND CLASS TICKETS,
AVAILABLE FOR ONE CALENDAR MONTH, ARE ISSUED FOR THIS TOUR

## (Via CHESTER, CARNARVON, MACHYNLLETH, AND SHREWSBURY, or *vice versa*.)

FROM THE UNDERMENTIONED STATIONS:

| STATIONS. | Snowdon Tour. | | STATIONS. | Snowdon Tour. | | STATIONS. | Snowdon Tour. | |
|---|---|---|---|---|---|---|---|---|
| | 1 Cl. | 2 Cl. | | 1 Cl. | 2 Cl. | | 1 Cl. | 2 Cl. |
| FROM | s. d. | s. d. | FROM | s. d. | s. d. | FROM | s. d. | s. d. |
| HITCHIN | 78 0 | 59 0 | Melton | 64 0 | 48 6 | Rotherham | 60 6 | 46 0 |
| BEDFORD | 73 0 | 55 0 | Lincoln | 66 6 | 51 0 | Loughboro' | 58 0 | 44 0 |
| WELLINGBORO' | 68 0 | 51 0 | Newark | 61 6 | 47 0 | Leicester | 60 0 | 47 0 |
| Northampton | 68 0 | 52 0 | Nottingham | 61 0 | 46 0 | Bristol | 98 0 | 58 0 |
| Cambridge | 80 0 | 62 0 | Trent | 55 0 | 42 0 | Bath | 78 0 | 58 0 |
| Huntingdon | 76 0 | 58 0 | Mansfield | 59 6 | 45 0 | Gloucester | 65 0 | 49 0 |
| Kettering | 67 6 | 51 6 | Chesterfield | 60 6 | 46 0 | Cheltenham | 65 0 | 49 0 |
| Market Harboro' | 64 0 | 49 0 | Masboro' | 60 6 | 46 0 | Tewkesbury | 65 0 | 49 0 |
| Peterboro' | 74 0 | 56 0 | Sheffield | 60 6 | 46 0 | Worcester | 57 0 | 44 0 |
| Stamford | 70 0 | 54 0 | | | | | | |

Passengers applying for these Tickets must state at the time of Booking which of the two undermentioned Tours they select :—
1st.—To travel via Carnarvon and thence to Barmouth or Dolgelly, returning via Welshpool, and Shrewsbury, to the Station at which they originally booked
2nd.—Travel via Welshpool and Barmouth to Carnarvon, returning thence by the Chester and Holyhead Line to the station at which they originally booked
  *From Leicester these Tickets are first-class through only, by those coaches*.
The Circular Tour is by Rail throughout. Coaches are run from Portmadoc and Carnarvon to Beddgelert, but the fares do not include conveyance by these coaches.
Omnibuses will also be run from Tanybwlch and Maentwrog to Penrhyndeudraeth. Holders of these tickets are at liberty to break their journey at any Station between Welshpool and Carnarvon via Bala, or at Bangor, Conway, Abergele, Rhyl, and Chester, or at any intermediate Station. Passengers holding these Tickets will have to provide their own conveyance from one Station to the other in Carnarvon.

ROUTES.—From Hitchin and Stations to Melton inclusive, Passengers travel by Uttoxeter, Crewe, and Carnarvon, returning via Welshpool as above, and proceeding via Stafford, Hinckley, and Stations to Loughboro' inclusive, via Uttoxeter, Crewe and Carnarvon, returning via Welshpool as above and proceeding via Stafford, and Ashburn or Tamworth, or vice versa.

From Lincoln and Stations to Loughboro' inclusive. Passengers travel via Birmingham, Stafford, Crewe and Carnarvon, returning via Welshpool as above, and proceeding via Stafford and Birmingham, or vice versa.
From Bristol and Stations to Worcester inclusive, Passengers travel via Birmingham, Stafford, Crewe and Carnarvon, returning via Welshpool, as above, and proceeding via Stafford and Birmingham, or vice versa.

*For General Conditions, see page 2.*

**All Trains over the Midland System convey First, Second, and Third Class Passengers.**

Rail and Steamer fares to Ilfracombe: Midland Railway handbill 16th May, 1874

the dock would be sufficiently completed for water to be let in on 1st May. The dock wall, 1,800 feet in length and tapering from a 28-foot width at its base to 7 feet at the top was finished. Disaster struck on 18th March, 1878 when just before 8.00 am, and fortunately just before the workmen arrived, a 90-foot length of wall collapsed due to the force of mud placed behind it as backing. The fall damaged a further 200 feet of wall. It took some time to recover the fallen masonry from the slushy mud and rebuild the wall. Charles Eckersley Daniel came on the scene but it is not clear whether he replaced Barnett & Gale or was a sub-contractor. *The Engineer* of 15th August, 1879 advertised that four standard gauge locomotives used in the dock construction were for sale by auction on 16th September, 1879 on behalf of C. E. Daniel.

The half-yearly meeting was to have been held on 28th February, 1879 but there was no quorum. It was announced that severe and continuous frost had delayed works on the dock wall, but that the coping would be set in two weeks.

The entrance to the dock was 560-foot long, 66-foot wide and 34-foot deep at high water. The wharf wall in a straight line ran for 1,800 feet and the 12½-acre dock had a depth of 24 feet. It was adjacent to a good anchorage ground at King Road.

The first ship to enter the dock was the Bristol & Portishead Railway Company's steamer *Lyn* on 28th June, 1879, the dock on the opposite bank at Avonmouth having opened on 24th February, 1877. The first commercial load arrived at Portishead Dock on 6th July, 1879 with the SS *Magdeburg* bringing in excess of 1,000 tons of barley.

# Chapter Five

## The Railway Evolves

At the half-yearly meeting in February 1874 Lewis Fry took the chair, Chairman James Ford being unwell and had to miss a meeting for the first time. It was announced that traffic had continued to improve, 147,639 passengers, excluding season ticket holders, being carried in the half-year against 131,876 the previous year; revenue was £5,574 against £5,034, while expenditure had decreased from £4,258 to £4,092; train miles were 23,713 against 22,135. As the original iron rails wore out they were replaced by those of steel, their cost included in the permanent way maintenance charged against revenue.

The Directors proudly announced to shareholders that no passenger had been injured in the seven years that the line had been opened and an estimated two million had been carried. From its opening the single line was worked by train staff and telegraph on the absolute block system, the signals were interlocked and the line examined daily. Safety was considered paramount and at all subsequent meetings the Directors proudly proclaimed that no passenger had been killed or injured. A less pleasant announcement was that no dividend would be declared until the dock was completed.

At the half-yearly meeting in August 1874 it was announced that 115,939 passengers had been carried against 105,560 of the previous comparable half-year, with receipts of £4,444 0s. 5d. against £4,249. Goods and mineral traffic increased train mileage from 19,658 to 22,162. Although goods and mineral traffic had increased, so had maintenance costs. Only one shareholder had not paid the demanded call on shares, so there was no problem with payment for the contractor.

In 1874 the Portishead branch was used for an interesting experiment. At that time there was no reliable method whereby a passenger could communicate with the driver if an emergency arose. The first attempt was made in 1847 when the Great Western Railway started fitting a hooded seat to the rear of locomotive tenders for a travelling porter whose task it was to look back along the train and inform the driver of any problem. This porter was popularly known as 'the man in the iron coffin'. In 1864 another method was tried. A cord was run from the front guard's van of a passenger train to a large gong mounted on the side of the tender, so that if necessary, he could raise an alarm. By 1869 this system had been modified and improved so that the cord extended through rings along the eaves of all the coach roofs and connected to the emergency whistle on the engine, instead of to the gong. This allowed anyone on the train, as long as they were able to lean out of a window and pull the cord, to warn the driver. It was not an ideal solution because sometimes a passenger had to haul in yards of loose cord until it was taut enough to sound the alarm – and in an emergency time was often of the essence.

Section 22 of the Regulation of Railways Act of 1868 required every passenger train travelling more than 20 miles without stopping, to be provided with an efficient means of communication.

On 23rd June, 1874 a Bristolian, Reuben Lyon, used the Bristol to Portishead line to test a further development. It consisted of a bellows and handle fixed to each compartment of a set of three carriages. By pulling down the handle, air was blown from the bellows through a pipe to both the guard's van and the engine, where it sounded whistles inserted into the ends of the pipes. In addition, the handle raised an arm, (or switched on a light at night), to indicate the compartment in which the apparatus had been used. The arm was cunningly placed so that it could not be lowered while the train was moving. The pipes doubled as a speaking tube between a guard and the driver. The inventor tried it with 300 feet of tube, equal to 10 or 12 coaches, and the six whistles invariably worked. The system was not adopted as it was more expensive to install than a simple cord and did not have any great advantage.

In November 1875 a 'communicator' was tried on the Portishead line – an electric bell which passengers could use to attract a driver's attention. By 1877 electrical systems had been adopted by some companies, but the best solution came as a result of the Regulation of Railways Act of 1889 requiring all passenger trains to be fitted with continuous brakes. It was relatively simple to fit a device whereby in an emergency a passenger could apply a train's vacuum or air brakes by means of a chain or handle. The 1868 Act allowed a penalty of £5 to be imposed for malicious activation.

At the August 1874 meeting the directors announced that goods and mineral traffic was increasing.

At the half-yearly meeting on 28th February, 1875 Daniel announced that in common with other railways, due to a depression in trade, a fall of passenger and goods receipts had been experienced: passenger numbers were 143,056 against 147,639; receipts £5,395 against £5,574 and expenses £4,430 against £4,092. As the stations had been in use for eight years, they had been renovated internally and externally while more steel rails had been laid.

As the half-yearly meeting held on 28th August, 1875 coincided with a meeting of the British Association in Bristol this resulted in there being insufficient shareholders present to form a quorum as on that same day the Association was visiting the dock works at Portishead. The weather was wet and the visitors appreciated the room where the railway directors

THE RAILWAY EVOLVES 33

GWR AEC lorry No. 411 registration LT 9967 in Cliff House Road, adjacent to Ashton Gate bridge, the wall of which can be seen on the far right. The picture was taken in 1932 when the vehicle had just killed a baby in a pram. *Author's collection*

Two 0-6-0PTs head a down goods through Clifton Bridge. One engine will probably spend the rest of the day shunting at Portishead. *Author's collection*

had provided luncheon. Passenger figures for the half-year were 6,927 first; 4,965 second and 67, 317 third, showing an increase of ½ per cent on first, 6 per cent on second and a decrease of 1¼ per cent on third, giving a revenue of £4,635 against £4,444 of the previous comparable half-year.

At the shareholders' meeting on 29th February, 1876 use of the railway had continued to increase: train mileage was 25,863 against 23,161; passenger numbers 148,137 against 148,055; receipts were £5,647 against £5,395 and season ticket sales had increased by 50 per cent.

On 30th August, 1876 due to the depression in trade figures had fallen: passenger numbers were 91,207 against 115,039; receipts £4,368 against £4,635 and mileage 22,716. Due to the amalgamation of the Bristol & Exeter with the Great Western, from 1st August, 1876 the line had been worked by the GWR. The chairman, James Ford, explained that the working agreement under the Act of 1863 was not liked by either the Portishead company or the Bristol & Exeter, so the latter had worked the line at cost, but this temporary arrangement had now stopped. With the amalgamation Act of 1874, the Portishead Company was to be paid 50 per cent of gross receipts instead of the actual working cost. The GWR maintained the line in addition to working it.

On 28th February, 1877 it was announced that there had been a small increase in passenger traffic and pier receipts, but a fall in mineral traffic due to the trade depression. When the demand was high for Welsh coal, that for gas works and local needs was brought by rail, but when the demand for Welsh coal slackened, most supplied arrived by sea.

At the meeting on 28th August, 1877 figures released showed that the usual increase in regular traffic had been maintained, but that due to the wet weather, excursion and tourist traffic for May and June had been below average; yet 124,319 passengers had been carried against 91,206 with a revenue of £4,632 against £4,308 and working expenses of £2,732 against £3,678. Revenue, instead of increasing as before in the spring and summer, had increased in winter showing 'it came from a more solid and permanent source'. Of an increase of 13,000, 11,000 had been in the winter months. It was realized that the gauge would have to be narrowed, but the Directors did not consider laying mixed gauge.

In the second half of 1878 the line carried 146,352 passengers with receipts being £5,539, both figures lower than the corresponding figures for 1877.

Due to a blizzard on 20th January, 1881 the first two trains to Portishead were cancelled and an irregular service run between 9 am and 9 pm. Between Pill and Portbury there were drifts up to ½-mile in extent. The branch was largely unaffected by the next Great Blizzard on 10th March, 1891.

Chapter Six

## The Line is Purchased by the Great Western Railway

The half-yearly report for the period ending 31st December, 1881 revealed that 165,882 passengers were carried by the railway; 32 steamers averaging almost 900 tons each and 14 sailing vessels entered the dock, the yearly total for the whole of 1881 being 60 steamers and 16 sailing vessels.

The opening of docks at Avonmouth and Portishead impinged on trade at Bristol Docks, so in November 1883 the Mayor of Bristol, Mr Weston, investigated the financial position of both the Avonmouth and Portishead companies and advised the corporation to promote a Bill to purchase both docks. The Bristol & Portishead Pier & Railway Company refused to sell the dock by itself and insisted that the railway and pier also had to be included in the sale.

The Great Western Railway (GWR), which had absorbed the Bristol & Exeter in 1876 was brought into the discussions and as from 1st July, 1884 the Portishead's company's assets – the railway and pier – were leased to the GWR by Act of 1884 47 & 48 Vict. cap. 256, 14th August, 1884, for an annual charge of £11,750. In consideration of a payment of £91,848 12s. 6d. in December 1891 the rent charge was abolished. The dock and warehouses at Portishead were taken over by Bristol Corporation.

Bristol Corporation's three docks specialized to a certain extent: Avonmouth concentrated on foreign trade; the city docks on coastal and near-continental trade, while Portishead also specialized on coastal trade.

Mercifully the branch had no serious accident, the worst being on Bank Holiday Monday 3rd August, 1885 when the 12.45 pm boat train from Temple Meads struck the buffers at Portishead Pier station. Consisting of a tank engine running bunker-first hauling 10 coaches fitted with continuous brakes and an unfitted luggage van, it had left Temple Meads 31 minutes late. It lost a further minute en route and on arrival at Portishead station, it ran round to propel the train to the Pier station. As was the practice, the station master travelled at what had become the front of the train, jumped off as it came to the Pier station and guided the driver by hand signals. Unfortunately he did not succeed in stopping the train before it collided with the buffers. Two passengers were cut on the forehead and three others shaken. A signalman riding in a brake van in the centre of the train received slight bruises to his face. The driver was said to be of excellent character, had been 38 years in the company's service, 33½ as driver, and never previously been involved in an accident. He had been on duty since 8.15 am.

In his Report, the Board of Trade Inspector, Colonel F. H. Rich, stated that a train with passengers should not be propelled into a station, particularly a terminus where there were facing points to pass over and buffer stops to butt against. He added that the yard between the ordinary passenger station and the pier needed rearranging and re-signalling; the points and signals should be interlocked and worked from raised signal cabins.

The GWR steamer services were withdrawn from Portishead Pier on 1st October, 1886, but Messrs P. & A. Campbell's paddle steamers continued to use it. The dock itself also was unprofitable and in 1886 made a loss in excess of £2,600 and by 1902 the loss totalled almost £160,000.

In the second half of the 19th century between the Bristol & Exeter's line at Portishead Junction and Clifton Bridge station, the Ashton Vale Iron Company had collieries and blast furnaces, both coal and iron producing traffic for the railway.

The Avon Gorge also had industry. Under an agreement of 1st October, 1891 the GWR agreed to the United Alkali Company making a subway under the Portishead branch between No. 2 and No. 3 tunnels in order to create an outlet from Greenland Quarry to a jetty on the River Avon. The work was carried out by the spring of 1892.

Bristol Corporation' Docks Committee, unhappy at the loss being made at Portishead sought a report from the engineer Wolfe Barry. This appeared in January 1896 and suggested three developments:

1. The dockization of the River Avon created by installing gates at the river mouth.
2  Making a second dock at Avonmouth.
3. Extending Portishead Dock into Portbury Marshes.

The second option was selected in 1900 and the necessary Act of Parliament secured on 17th August, 1901, the Royal Edward Dock, Avonmouth being opened on 9th July, 1908.

The GWR was unhappy at this development and would have preferred the money to be spent on development at Portishead where it had the monopoly, whereas at Avonmouth it had to share traffic with the Midland Railway. However Portishead was not entirely neglected and in 1903 the Docks Committee agreed to construct a timber wharf on the Portbury side of Portishead Dock. There was to be 200-foot long jetty and 600-foot long timber wharf with a 10-acre stacking ground for

## Portishead Flower Show and Fete.

### On MONDAY, AUGUST 3rd,
EXCURSION TICKETS WILL BE ISSUED TO

# PORTISHEAD

| FROM | By Trains as under: | | | | | | | Return Fares, 3rd Class. | |
|---|---|---|---|---|---|---|---|---|---|
| | | | | | | | | To Portishead and back. | To Portishead, returning from Clevedon. |
| | A.M. | A.M. | A.M. | P.M. | P.M. | P.M. | P.M. | s. d. | s. d. |
| Clifton Down | 8 57 | 9 38 | 10 37 | .. | .. | 12 59 | 1 39 | } 1 4 | 1 9 |
| Redland .. .. | 9 0 | 9 40 | 10 39 | .. | .. | 1 1 | 1 41 | | |
| Montpelier .. | 9 3 | 9 42 | 10 41 | .. | .. | 1 4 | 1 43 | } 1 4 | 1 8 |
| Ashley Hill .. | 8 58 | .. | 10 50 | .. | .. | 12 24 | .. | | |
| Stapleton Road | 9 6 | 9 47 | 10 56 | .. | .. | 1 10 | 1 47 | } 1 3 | 1 6 |
| Lawrence Hill | 9 10 | 9 50 | 11 2 | .. | .. | 1 15 | 1 49 | | |
| St. Anne's Park | .. | 9 26 | 10 57 | .. | .. | .. | 2 5 | | |
| Temple Meads | 9 20 | 10 30 | 11 15 | .. | .. | .. | 2 15 | | |
| Bedminster .. | 9 27 | 10 35 | 11 20 | 12 2 | .. | 1 30 | 2 20 | 1 2 | 1 4 |
| Ashton Gate Platform | 9 33 | 10 42 | 11 29 | 12 6 | 1 0 | 1 36 | 2 25 | 3 38 | 1 1 | } 1 * 4 |
| Clifton Bridge* | 9 35 | 10 45 | 11 30 | 12 10 | 1 2 | 1 38 | 2 30 | 3 40 | 1 0 | |

TIMES OF RETURN TRAINS.—From Portishead to St. Anne's Park at 5.20 and 7.35 p.m.; to Ashley Hill at 5.20, 8.15 and 9.45 p.m.; and to other stations at 5.20, 6.25, 6†45, 7†10, 7.35, 8.15, 8†45, 9.45 and 10.10 p.m.
                    † For Clifton Bridge and Ashton Gate only.
Passengers holding Ashley Hill tickets may return to Montpelier if they so elect.
N.B.—Passengers holding tickets to Portishead, available to return from Clevedon, can only travel by the return excursion trains from Clevedon shewn below.
* Tickets issued at Clifton Bridge and Ashton Gate, available for returning from Clevedon, will be available from Clevedon to Bedminster only on return journey.
N.B.—Tickets for Portishead are not issued at Messrs. Cook & Son's Office, 49 Corn Street, "Pickford's" Office, St. Augustine's Parade (Tramway Centre), or at Mr. H. W. Gapper's Office, 2 Cromwell Road (Zetland Road Junction), Montpelier.

### On MONDAY, AUGUST 3rd,
EXCURSION TRAINS will run from BRISTOL to

# Clevedon & Weston-s.-Mare

| Leaving | STARTING TIMES. | | | | | | | | | | |
|---|---|---|---|---|---|---|---|---|---|---|---|
| | A.M. | A.M. | A.M. | A.M. | A.M. | A.M. | A.M. | A.M. | A.M. | NOON | NOON | P.M. |
| Avonmouth Dock | ... | 8 2 | 9 0 | ... | ... | 10 0 | ... | 10 50 | ... | 11 40 | 11 55 | 12 38 |
| Shirehampton ... | ... | 8 7 | 9 5 | ... | ... | 10 5 | ... | 10 55 | ... | 11 44 | 12 0 | 12 42 |
| Clifton Down ... | 7 43 | 8 38 | 9 20 | 9 38 | 10 10 | 10 37 | ... | 11 7 | 11 30 | 12 15 | 12 36 | 12 58 |
| Redland... ... ... | 7 45 | 8 41 | 9 22 | 9 40 | 10 12 | 10 39 | ... | 11 9 | 11 32 | 12 17 | 12 38 | 1 0 |
| Montpelier ... ... | 7 47 | 8 44 | 9 24 | 9 42 | 10 15 | 10 41 | ... | 11 11 | 11 34 | 12 19 | 12 40 | 1 3 |
| Ashley Hill ... | ... | ... | 9 28 | 9 45 | 10 20 | 10 50 | ... | ... | 11 48 | 12 0 | ... | 1 12 |
| Stapleton Road ... | 8 2 | 8 55 | 9 35 | 10 10 | 10 25 | 10 55 | 11 5 | 11 55 | 12 32 | 12 58 | 1 15 |
| Lawrence Hill ... | 8 8 | 9 0 | 9 40 | 10 15 | 10 30 | 11 0 | 11 12 | 11 36 | 12 0 | 12 38 | 1 5 | 1 20 |
| St. Anne's Park ... | ... | ... | ... | 9†26 | ... | 10*33 | ... | 11*5 | 11*24 | ... | 12*25 | ... |
| Bedminster ... | 8 20 | 9 15 | 9 54 | 10 28 | 10 45 | 11 10 | ... | 11 42 | 12 16 | 12 50 | 1 22 | 1 40 |

| | P.M. | P.M. | P.M. | TIMES OF RETURN TRAINS. |
|---|---|---|---|---|
| Avonmouth Dock | 1 22 | 1 42 | 1 53 | From WESTON (Excursion Platform) to Shirehampton and Avonmouth Dock at 5.40, 7.20, 8.15, and 8.25 p.m., to St. Anne's Park at 7*20, 8*5, and 9*0 p.m., to Ashley Hill at 6.50, 8.5 8.25, 9†25 and 10.15 p.m., to other Stations at 5.40, 6.50, 7.20, 8.5 p.m., and afterwards about every 20 minutes until 11.5 p.m. From CLEVEDON to Shirehampton and Avonmouth Dock at 5.35, 7†5, 8†10 and 8.50 p.m., to St. Anne's Park at 8*50 p.m., to Ashley Hill at 6.50, 8.50 and 9.35 p.m., and to other Stations at 5.35, 6.50, 7.5, 8.10, 8.50 and 9.35 p.m. |
| Shirehampton | 1 27 | 1 46 | 1 58 | |
| Clifton Down | 1 39 | 2 0 | 2 35 | |
| Redland | 1 41 | 2 2 | 2 37 | |
| Montpelier | 1 43 | 2 5 | 2 39 | |
| Ashley Hill | ... | ... | 2 55 | |
| Stapleton Road... | 1 56 | 2 15 | 3 0 | |
| Lawrence Hill | 2 3 | 2 20 | 3 5 | |
| St. Anne's Park | 1*54 | ... | ... | |
| Bedminster | ... | 2 35 | 3 18 | |

    * Change at Lawrence Hill.    † Change at Temple Meads.    ‡ Change at Bedminster.
Passengers holding Ashley Hill tickets may return to Montpelier if they so elect.
[Continued on page 79.

For Special Notices relating to the issue of Excursion Tickets, see page 2.

Handbill advertising Portishead Flower Show & Fete 1908

View from the GWR at Portishead towards the Weston, Clevedon & Portishead Light Railway station. The link between the two railways can be seen in the bottom left-hand corner.
*Author's collection*

Ex-London Brighton & South Coast Railway's 'Terrier' 0-6-0T *Portishead* at the Weston, Clevedon & Portishead Light Railway station, Portishead. The wagons on the right are going to or from the GWR.
*Author's collection*

storing between 20,000-30,000 standards of timber, each standard containing approximately $3\frac{1}{3}$ tons of wood. Much of the timber was taken away by rail.

A further development at Portishead was in June 1908 when British Petroleum, on behalf of the Anglo-Saxon Petroleum Company, leased about three acres of land on the Portbury side of the lock at Portishead. This was used for erecting storage tanks for fuel brought by sea. With the development of the internal combustion engine for road transport, by 1911 the area had grown to six acres and almost seven acres by 1913. A small refinery was built to treat the crude oil while a nest of sidings was provided for rail tanks. The siding serving these facilities was an extension of the timber siding.

As Timber Jetty ground frame giving access to these sidings was on a single line, when they needed to be shunted the single line staff had to be obtained from Portishead signal box, and in order not to block the line between Portishead and Pill, the locomotive was required to be locked in the siding and the staff returned to the box by a porter who later had to collect the staff and take it to release the engine when shunting duties were completed.

On 7th August, 1907 the Portishead extension of the Weston, Clevedon & Portishead Light Railway (WCPLR) was opened (*see plan, page 97*). Exchange sidings under an agreement of 11th August, 1906 were laid and first used on 2nd November, 1908. When the light railway closed on 18th May, 1940, the sidings were retained.

Although the WCPLR had two petrol-engined railcars, most of the Clevedon to Portishead trains were steam-hauled so that en route they could call at the quarries and attach up to about 10 loaded wagons behind the passenger carriages, a total of 40 or more wagons being dealt with on a busy day. In 1934 an average of 15 wagons daily ran from Black Rock and Conygar quarries. Stone from Conygar tended to be destined for Toton and that from Black Rock to Acton and Hayes. Clean crushed stone travelled in common user wagons, or GWR ballast hoppers if the stone was destined for that company. Steel-lined wagons were used for tarred roadstone. At Portishead the wagons were placed on the weighbridge and then propelled to the GWR exchange sidings 22 chains beyond the WCPLR's station at Portishead.

Similarly Portishead to Clevedon trains took empty wagons to the quarry sidings in addition to tar and bitumen tanks and coal for liquifying these substances. Key's oil tank No. 10 regularly ran from the company's works at Portishead Docks to the quarry with residual oil for making asphalt.

Most of the traffic destined for the light railway was coal, mainly from the East Midlands collieries at Bilsthorpe, Blackwell and Shireoaks, with a smaller tonnage from Trentham, Staffordshire. Much of the coal was destined for Clevedon gas works, with a smaller proportion for merchants at Clevedon, Worle, Milton Road and Weston-super-Mare. Coal traffic was carried in private owners' wagons. Periodically fertiliser arrived consigned to various destinations. During the First World War a remount depot was established at Shirehampton on the Bristol to Avonmouth line, and some of the manure and used bedding was taken to Portishead and then onwards to Cadbury Road station in the Gordano Valley. With the closure of the WCPLR in 1940, until the 1950s the output of Black Rock Quarry arrived by road and was put on rail at Portishead.

In an effort to combat road competition, in August 1908 the GWR reduced some fares: Clifton Bridge to Portishead became 1s. rather than 1s. 4d. and Pill to Portishead 6d. instead of 8d.

On Bank Holiday Monday 3rd August, 1908 in the evening Portishead station was crowded with excursionists wanting to go home. With an estimated 2,000 persons on the platform, at 9.00 pm when a train was being propelled into the station, a child aged about four fell between the coaches and the line. Screams from passengers alerted railway employees who rescued him unhurt. Although several coaches passed over him, he was safely between the rails.

On 16th December, 1910 a serious flood covered land east of Portishead Dock, both the GWR and WCPLR stations were inaccessible as was the gas works, flooding of the latter caused the gas supply to be cut off.

In November 1912 the GWR Directors considered building a new joint passenger station at Portishead to facilitate transfer between the GWR and WCPLR, but the idea was not taken further. In 1954 a new station was built on the site proposed in 1912, but for a different reason as the WCPLR by then had closed.

On 18th August, 1911, a large number of business people who lived at Portishead and worked in Bristol, anticipating the Railway Strike disorganizing the train services, arrived to catch the first train, the 6.55 am, only to find that it had been cancelled. Two gentlemen, anxious to reach Bristol, hired the only taxicab of which the town could boast. The other passengers used their initiatives and hired a brake and pair from the local livery stable.

The gentlemen in the taxi felt less superior when after proceeding only a few hundred yards, it emitted an ominous noise and as the driver turned his head to endeavour to establish the cause, the taxi mounted the pavement and nearly overturned. It stopped and an inspection revealed that the driving chain had broken.

While the driver was attempting to make a repair, a private motorist came along and offered them a lift. This was gladly accepted and on their way to Bristol they met the horse brake returning! The motorists were also beaten to Bristol by a further group from Portishead who had used bicycles. Some decided to stay overnight in Bristol to avoid any problems the following morning.

The signal box at Portishead Junction was set at the foot of a deep cutting with the sloping bank crowned by a low wall bordering the Bridgwater Road. Very early on Sunday 20th August, 1911 a disorderly rabble gathered there, at first jeering at the signalman on duty for breaking the strike. When they started to throw stones, he telegraphed to Temple Meads for assistance.

In reply, soldiers on duty at Pylle Hill were dispatched by light engine. Temple Meads additionally contacted Bridewell police station and a posse of constables set off for Portishead Junction in taxis and fire brigade vans. Finding that the mob persisted in their attack on the signal box and the force for its protection was so small, the military, under orders, fired blank cartridges over the heads of the mob. This steadied them for a time, but when the police arrived, they were forced to charge at the rabble with batons. This action proved effective and order was restored. The police remained in the neighbourhood. The Chief Constable visited the scene and on his instruction, one of the fire engines accompanied the police. The Bristol City Marine Ambulance was called out and an unconscious man suffering from concussion was taken to the General Hospital in their wagon.

On the outbreak of war in 1914, the Port of Bristol Authority (PBA) had two locomotives shedded at Portishead. The war had its effect on trade: there was a decrease in imported grain and timber, but a very large increase in demand for petrol, Portishead becoming the main supply centre for the country. Petrol was placed in cans for easy distribution for use by road vehicles or aircraft. Imports over the four years totalled 116 million gallons and exports, mainly to France, totalled 72 million.

Every morning and afternoon in 1917 a 45-wagon train ran from Portishead to Rochester on the South Eastern & Chatham Railway via Reading, GWR locomotives running through from Portishead to Redhill. The morning train was headed by a 2-6-0 of the '43XX' or 'Aberdare' class and travelled via Badminton. The afternoon train hauled by a Dean Goods 0-6-0 travelled via Bath and regulations stipulated that the Box banking engine was required to be ready to assist.

A new product to Portishead was toluol (now called toluene), used in the manufacture of trinitrololuene (TNT) used for manufacturing bombs, mines and shells. Hitherto Britain had extracted it from coal, but crude petroleum could also be used. Messrs Shell offered a supply of toluene from Borneo, but there was no distillery available in Britain to separate toluene from the crude petroleum.

A solution was at hand. Messrs Shell had such a distillery in Rotterdam and they moved it in batches to Portishead. A site on the PBA's land was prepared and within six weeks the factory was in full production. The toluene was taken to armament factories by rail.

As many railwaymen had enlisted, coupled with the fact that wartime freight needed priority, from January 1917 Clifton Bridge, Pill, Portbury and Portishead stations were closed on Sundays and the same month a 50 per cent increase in fares was imposed to help deter passenger traffic and to assist in defraying increased costs due to the war.

One wartime problem was the replacement of ships destroyed by enemy action, so in July 1917 Parliament approved the establishment of the National Shipyard Company opening three new shipyards located at Beachley, Chepstow and Portbury and as these were all on the Bristol Channel, the hulls could all be conveniently fitted-out at Avonmouth. Portbury in addition to shipbuilding, was also the site for the organization's headquarters.

The actual shipyard was sited at Sheephouse Farm, Easton-in-Gordano and constructed by 1,800 men of the Royal Engineers, who were billeted in various homes at Pill, but later housed in a purpose-built camp. To aid construction, commencing on 5th August, 1918 loaded gravel trains began working from Frocester on the Midland Railway's Bristol-Gloucester line, to Portbury Shipyard, hauled throughout by Midland engines. A junction at 127 miles 18 chains was opened on 3rd October, 1917 to enable materials to be brought by rail. One engine used in the construction works was the Port of Bristol Authority's 0-6-0ST *Portbury*. A signal box and four sidings were brought into use on 29th January, 1918, and a halt opened at 127 miles 12 chains on 16th September, 1918. The signal box had electric token working to both Portishead and Pill, and was provided with a block switch. If switched out the long section Pill-Portishead continued to be worked by electric staff. Although the slipways had been laid down, the armistice meant that the project was abandoned, so Portbury Shipyard

PRIVATE AND NOT FOR PUBLICATION.　　　　　Notice No. 901.

# GREAT WESTERN RAILWAY.

**OPENING OF NEW STATION on the Portishead Branch on the Pill Side of Portbury Shipyard Exchange Sidings (mileage 127 m. 12½ chs.), to be named PORTBURY SHIPYARD PLATFORM, MONDAY, SEPTEMBER 16th, 1918.**

This New Station, having access from the public road at Sheep House Lane Overbridge between Pill and Portbury, will be opened on Monday, September 16th, 1918, and passenger trains will be timed to call there in accordance with Public Bill No. B.890, and the working Time Table shewn below.
The platform is situated on the Up side of the Portishead Branch single line.
Passenger and parcels traffic of all descriptions will be dealt with.
The Station will be under the supervision of the Station Master at Portbury, and the trains and traffic will be attended to by the checkers working in turn at Portbury Shipyard Exchange Sidings, assisted, as may be necessary, by other members of the Portbury Staff.
Booking Clerks are requested to take care that correct tickets are issued to persons travelling to Portbury Shipyard Platform and Portbury respectively.

## REVISED TIME TABLE OF PORTISHEAD BRANCH PASSENGER TRAINS.

### DOWN TRAINS—WEEK DAYS.

| Distance from Paddgtn. | | | Pass. | Pass. | Mtr. | Pass. | Pass. SO | Mtr. VV | Pass. | Pass. | E'ty C'chs SX | Mtr. | Frome Pass. | Pass. | Pass. SO |
|---|---|---|---|---|---|---|---|---|---|---|---|---|---|---|---|
| M | C | | A.M. | A.M. | A.M. | P.M. | P.M. | P.M. | P.M. | P.M. | P.M. | P.M. | P.M. | P.M. | P.M. |
| 118 | 28 | Bristol (T. Mds.) dep. | 7 0 | 8 20 | 9 50 | 12 15 | .. | 2 5 | 4 15 | 5 30 | .. | 6 30 | 7 50 | 9 25 | 11 0 |
| 119 | 32 | Bedminster .. ,, | 7 4 | 8 24 | 9 54 | 12 19 | .. | 2 9 | 4 19 | 5 34 | .. | 6 35 | 7 54 | 9 29 | — |
| 120 | 27 | Portishead Jct. dep. | 7 | 8 57 | 13 32 | | .. | 2 13 | 4 22 | 5 37 | .. | 6 38 | 7 57 | 9 32 | 11 5 |
| 121 | 62 | Clifton Bridge { arr. | 7 10 | 8 30 | 10 0 | 12 25 | .. | 2 16 | 4 25 | 5 40 | .. | 6 40 | 8 0 | 9 35 | 11 8 |
| | | { dep. | 7 14 | 8 32 | 10 2 | 12 27 | .. | 2 17 | 4 26 | 5x43 | 6†15 | 6x43 | 8 1 | 9x36 | 11 10 |
| 126 | 13 | Pill .. { arr. | 7x23 | 8 41 | 10 11 | 12 36 | .. | 2 26 | 4 35 | 5x52 | 6†25 | 6 52 | 8 10 | 9 45 | — |
| | | { dep. | 7 25 | 8x45 | 10 13 | 12 38 | .. | 2 28 | 4x37 | 5 54 | 6x34 | 6 53 | 8 12 | 9 47 | CS |
| 127 | 12½ | PortburyShip- { arr. | 7 29 | 8 49 | 10 17 | 12 42 | .. | 2 32 | 4 41 | 5 58 | — | 6 58 | 8 16 | 9 51 | 11 22 |
| | | yard Platform { dep. | 7 31 | 8 50 | 10 18 | 12 43 | 1 24 | 2 33 | 4 42 | 5 59 | — | 6 58 | 8 17 | 9 52 | 11 26 |
| 127 | 46 | Shipyard Sidings ,, | CS | CS | CS | CS | CS | CS | CS | CS | CS | CS | CS | CS | CS |
| 127 | 77 | Portbury { arr. | 7 33 | 8 52 | 10 20 | 12 45 | 1 26 | 2 35 | 4 44 | 6 1 | — | 7 0 | 8 19 | 9 54 | — |
| | | { dep. | 7 37 | 8 55 | 10 22 | 12 47 | 1 28 | 2 37 | 4 46 | 6 3 | — | 7 3 | 8 22 | 9 57 | — |
| 129 | 75 | Portishead .. arr. | 7 42 | 9 0 | 10 27 | 12 52 | 1 33 | 2 42 | 4 51 | 6 8 | 6†46 | 7 8 | 8 27 | 10 2 | 11 32 |

VV Will be train SO.

### UP TRAINS—WEEK DAYS.

| | | | Pass. | Pass. | Mtr. | E'ty C'chs SO₁ | Pass. | Mtr. VV | Pass. | Pass. | Pass. | Mtr. SX | Pass. | E'ty C'chs SO |
|---|---|---|---|---|---|---|---|---|---|---|---|---|---|---|
| | | | A.M. | A.M. | A.M. | P.M. | P.M. | P.M. | P.M. | P.M. | P.M. | P.M. | P.M. | P.M. |
| Portishead .. .. .. dep. | | | 7 10 | 8 30 | 10 45 | 12†55 | 1 10 | 3 5 | 5 15 | 5†35 | 6 18 | 7 20 | 9 10 | 11†45 | .. |
| Portbury .. .. ,, | | | 7 15 | 8 35 | 10 50 | — | 1 15 | 3 10 | 5 20 | — | 6 23 | 7 25 | 9 15 | — | .. |
| Shipyard Sidings .. ,, | | | CS | CS | CS | 1† 2 | CS | CS | CS | CS | CS | CS | CS | CS | .. |
| Portbury Shipyard { arr. | | | 7 17 | 8 37 | 10 52 | | 1 17 | 3 12 | 5 22 | 5†43 | 6 27 | 7 27 | 9 17 | — | .. |
| Platform .. { dep. | | | 7 19 | 8 38 | 10 53 | Q | 1 19 | 3 13 | 5 23 | 5 48 | 6 29 | 7 28 | 9 18 | — | .. |
| Pill .. .. .. { arr. | | | 7x23 | 8 42 | 10 57 | .. | 1 23 | 3 17 | 5 27 | 5x52 | 6x33 | 7 32 | 9 22 | CS | .. |
| { dep. | | | 7 25 | 8x43 | 10 58 | .. | 1 25 | 3 18 | 5 32 | 5 53 | 6 34 | 7 33 | 9 23 | — | .. |
| Clifton Bridge .. { arr. | | | 7 34 | 8 52 | 11 7 | .. | 1 34 | 3 27 | 5x41 | 6 1 | 6x41 | 7 42 | 9x32 | — | .. |
| { dep. | | | 7 35 | 8 54 | 11 8 | .. | 1 35 | 3 29 | 5 43 | — | 6 44 | 7 44 | 9 34 | CS | .. |
| Portishead Jct. .. arr. | | | 7 38 | 8 58 | 11 12 | .. | 1 39 | 3 33 | 5 47 | .. | 6 48 | 7 48 | 9 37 | 12 10 | .. |
| Bedminster.. .. { arr. | | | 7 41 | 9 1 | 11 15 | .. | 1 42 | 3 36 | 5 50 | .. | 6 51 | 7 51 | 9 40 | — | .. |
| { dep. | | | 7 44 | 9 3 | 11 18 | .. | 1 45 | 3 38 | 5 52 | .. | 6 53 | 7 54 | 9 44 | — | .. |
| Bristol (T. Meads) arr. | | | 7 48 | 9 7 | 11 22 | .. | 1 49 | 3 42 | 5 57 | .. | 6 57 | 7 58 | 9 48 | 12†14 | .. |

Q To be run round in Exchange Sidings and backed to Shipyard Platform.　　VV Will be train SO.

**H. R. GRIFFITHS,**
September 12th, 1918.　　　　　*Superintendent of Bristol Division.*

(700 R. 8vo.)　　J. W. Arrowsmith Ltd., Quay Street, Bristol.

Halt closed 26th March, 1923, though on 2nd April, 1928 one of the sidings became a crossing loop for passenger trains. One siding was removed and the two remaining sidings each held 44 wagons. Over £11,600 had been spent on the railway works for Portbury Shipyard plus about £7,300 on the laying sidings at Clifton Bridge and West Depot.

The shipyard site became Central Stores Depot No. 367 for the disposal of surplus equipment not required by the services in peacetime and the plant was sold there on 31st October, 1926. The shipyard site is now part of the Royal Portbury Dock.

On 28th June, 1927 the GWR using one of its forward-control Burfords, inaugurated a bus service between Portishead station and the Nautical School at Redcliffe Bay, running until 19th July, 1923 when it was taken over by Bristol Tramways & Carriage Company. On 31st December, 1928 the GWR ran an experimental bus service Mondays-only, from Portishead to Clevedon, Yatton and Claverham. Traffic was light and 1st July, 1929 was the last day of its operation.

**WEEKLY RESIDENTIAL TICKETS**

ARE ISSUED BETWEEN BRISTOL STATIONS AND

**Portishead, Clevedon & Weston-super-Mare,**

Entitling the holder to travel by any of the Company's Passenger Trains between the points named, for one week.

| Return Fares, to or from | Portishead. 1st cl. s. | d. | Portishead. 2nd cl s. | d. | Clevedon. 1st cl. s. | d. | Clevedon. 2nd cl s. | d. | Westn.-s/M 1st cl. s. | d. | Westn.-s/M 2nd cl s. | d. |
|---|---|---|---|---|---|---|---|---|---|---|---|---|
| Clifton Down | 13 | 3 | 10 | 0 | 16 | 3 | 11 | 6 | 18 | 0 | 12 | 6 |
| Redland .. .. | 13 | 0 | 9 | 9 | 16 | 9 | 11 | 0 | 17 | 9 | 12 | 3 |
| Montpelier .. | 12 | 9 | 9 | 6 | 15 | 6 | 10 | 9 | 17 | 6 | 12 | 0 |
| Ashley Hill .. | 12 | 3 | 9 | 3 | 15 | 0 | 10 | 6 | 17 | 0 | 11 | 9 |
| Stapleton Rd. | 11 | 6 | 8 | 9 | 14 | 6 | 10 | 3 | 16 | 6 | 11 | 6 |
| Lawrence Hill | 11 | 0 | 8 | 6 | 13 | 9 | 9 | 9 | 16 | 9 | 11 | 0 |
| Temple Meads | 10 | 6 | 8 | 0 | 13 | 9 | 9 | 0 | 15 | 0 | 10 | 6 |
| Bedminster .. | 10 | 0 | 7 | 9 | 12 | 0 | 8 | 6 | 14 | 3 | 10 | 0 |

N.B.—Tickets to Portishead, Clevedon and Weston-super-Mare are not issued at Messrs. Cook and Son's Office. 49 Corn Street, at "Pickfords," St. Augustine's Parade (Tramway Centre), or at Mr. H. W. Gapper's Office, 2 Cromwell Road (Zetland Road Jct.), Montpelier.

Section from a handbill advertising Weekly Residential Tickets 1908

# Chapter Seven

## Developments

On 27th July, 1918 an agreement was signed between the Imperial Tobacco Company and the GWR to provide two looped sidings each holding 20 wagons serving the firm's premises near milepost 121, they came into use 7th November, 1918. There was a private siding agreement 12th December, 1919 with the adjacent Ashton Saw Mills Ltd and 23rd July, 1937 with Ashton Containers Ltd.

On 22nd December, 1926 1½ miles beyond Clifton Bridge, as the 6.28 pm from Temple Meads was passing through the Avon Gorge the front axle of the locomotive broke. The driver stopped and no coaches were derailed. A relief train was sent from Portishead to collect the passengers – mostly businessmen and Christmas shoppers, who were considerably delayed. Some, rather than wait, walked back to Clifton Bridge station and either caught a bus or hitched a ride in a car. The passenger service continued throughout the evening, passengers walking past the derailment.

The Electrical Commissioners under the 1926 Electricity Supply Act decided to build Portishead 'A' Power Station. Messrs William Cowlin won the building contract. On the standard gauge Cowlin's used ex-Weston, Clevedon & Portishead Railway Manning, Wardle 0-6-0ST *Portishead*, while on the 2-foot gauge they ran a 0-4-0ST Kerr, Stuart

Weston, Clevedon & Portishead Light Railway Manning, Wardle 0-6-0ST *Portishead* engaged on the power station project in 1927. *M. J. Tozer collection*

No. 3090 of 1917 which on completion of the contract was transferred to Keynsham for building Fry's Somerdale works.

The 'A' power station opened on 25th March, 1927 and the associated sidings came into use on 1st February, 1929. Some coal came by sea from South Wales while the remainder came by rail from North Somerset. At 129 miles 32 chains – the Bristol end of the rail complex at Portishead – on 14th November, 1927 a private siding agreement was made with Severn Kraft Mills, the lines coming into use on 13th February, 1928. In June 1934 the firm went into compulsory liquidation, but the sidings remained.

In 1929 in preparation for the inauguration of a half-hourly service during rush hours, a loop at Oakwood was laid just east of Ham Green Halt. Although most trains crossed at Pill or Portbury Shipyard, the loop at Oakwood permitted crossing if trains were running out of course.

To cope with the increased service, at Portishead anup platform was built beside the former carriage siding which became a passenger loop, this came into use on 9th March, 1930. Both platforms were signalled to receive and dispatch trains.

To help ease unemployment during the Depression, in 1929 the Government offered loans to carry out large public works and the extension of Temple Meads station qualified. P. E. Culverhouse drew plans to more than double the size of the station and provide all the main platforms with refreshment and waiting rooms.

To the north of Temple Meads the track was quadrupled between Filton Junction and Stapleton Road, Dr Day's Bridge Junction and Temple Meads and from Temple Meads to Portishead Junction, a distance overall of nearly seven miles. Between Temple Meads and Portishead Junction, Bedminster and Parson Street stations needed rebuilding to cope with the quadrupling.

West of Parson Street station, was a 56 ft skew arch which carried Bedminster Down Road over the railway and when rebuilt, the two skew faces were formed in concrete, strengthened with steel bars before the middle section was built in brick. The original bridge was demolished on Sunday 26th April, 1931 and the one at Parson Street on 15th November, 1931. On both occasions passenger trains were diverted running via the Harbour line through Wapping, Ashton Swing Bridge to West Depot.

During the Second World War on 27th June, 1940 seven high-explosive bombs made craters at the edge of the railway between Portbury and Portishead, only slightly damaging track and telegraph wires. On 2nd December, 1940 a bomb fell near Oak Wood signal box, an unexploded bomb fell over the tunnel at Ham Green and another unexploded bomb fell near Clifton Bridge. The line was re-opened on 7th December but owing to problems at Temple Meads, services were only run from Clifton Bridge-Portishead. On 18th January, 1940 there was a crater and debris between Pill and Portbury, but the line was cleared by 8.00 am. On 26th June, 1940 a bomb dropped beside the line at Portbury the General Manager noting in his monthly report to the Board that it only caused 'trifling damage'. On 3rd-4th April, 1941 telegraph wires were brought down between Clifton Bridge and Portishead while on 17th April, 1941 an unexploded bomb at Ashton Gate caused the suspension of services. Although high explosive bombs had fallen on the pier lines at Portishead, damage did not interfere with station activities. On 11th-12th May, 1942 four high-explosive bombs fell near one of the Portishead branch tunnels causing communication failure requiring pilotman working, though with suspension from 11.26 am–3.30 pm owing to a suspected unexploded bomb.

On 24th-25th November, 1940 when the main line between Temple Meads and Parson Street had been bombed in several places, it is believed that some through main line trains were diverted over the Harbour line and Ashton Gate.

In 1949 the oil-burning Portishead 'B' power station was authorized. The 26½ acre site lay between the entrance lock to the dock and the GWR railway station. As the station site was required for access, a new station

Albright & Wilson's Peckett 0-4-0ST Works No. 1611, 22nd May, 1964.    *Revd Alan Newman*

was built about half a mile to the south and opened on 4th January, 1954. The transit sheds and the PBA granary also required demolition.

The needs of the generating stations demanded seven colliers in and out on every tide. The coal was handled mechanically. Between 1971 and 1974 the boilers were converted to oil-burning and the last coal train ran in 1974.

Much of the current the new power station generated was used by Albright & Wilson, the chemical firm on the far side of the lock. It was well-sited, phosphate arriving by sea, the other constituents needed, anthracite and silica pebbles from Budleigh Salterton, arrived by train. The first load of raw materials arrived on 8th October, 1953 and the first rail tanks of phosphorous left on 22nd February, 1954. The 24 specially-built wagons held a payload of 22 ¼ tons each. As phosphorous was liable to spontaneous combustion if in contact with air, it was loaded and discharged in water at a temperature of 60 degrees Celsius. Full tank wagons travelled with a water 'blanket' and the tanks were filled with water for the return journey both on safety grounds and to protect the tank from internal attack by acids formed from residues remaining in the tank. At the end of December 1956 the first load of phosphates was taken by rail to the firm's Oldbury Works, Birmingham. Traffic averaged 36-40 tankers weekly. The liquid phosphorous needed careful handing and leaking valves on the tanks could cause sleepers to ignite. As Albright & Wilson's at Portishead was being developed, so both it and 'B' power station received a large mount of steel for construction. Around 1964 phosphate trains used BR plastic wagon sheets replacing the canvas tarpaulins.

Albright & Wilson had prototype Peckett diesel Works No. 5000 on loan 14th August, 1956.
*Revd Alan Newman*

Albright & Wilson's Ruston & Hornsby diesel-electric 0-4-0 Works No. 38175, 14th February, 1972.
*Revd Alan Newman*

On 24th August, 1955 a private siding agreement was signed with the Ministry of Fuel & Power for six sidings for Esso to be laid to provide oil storage in an emergency. The agreement was terminated on 31st December, 1967 and the roads lifted by June 1969.

In the 1950s approximately 150,000 tons of timber arrived annually. Slings of timber were unloaded by dock steam cranes from a ship to either road or rail wagons, or were taken by crane to the stacking ground for storage.

On Sunday 3rd February, 1957 Flying Officer Crossley in a Vampire jet took off from Filton airfield shortly before the disbanding parade of the 501 Squadron based there. He dived below Clifton Suspension Bridge, made a victory roll and struck a steep slope causing a slight landslide which threatened the Portishead line, so traffic was suspended for a short period until it was established that it was safe to carry traffic. Crossley did not survive the crash.

In the 1970s Albright & Wilson's wagons were becoming life expired so there were two options: purchase new freight tanks, or change to road transport. As the replacement rolling stock would have to be bogie vehicles and the curves on the factory lines were too tight, the latter option was selected. This affected BR staff at Portishead for as dryness was essential, all loads had to be sheeted, or sometimes, double-sheeted and often this was carried out 12 hours a day, seven days a week, so this work was lost.

On 9th September, 1958 Portishead Docks were temporarily closed to enable the necessary repairs to be effected to the entrance lock. This impinged on the Central Electricity Generating Board (CEGB) which

# PUBLIC NOTICE
## BRITISH RAILWAYS BOARD (WESTERN REGION)

### TRANSPORT ACT 1962
### WITHDRAWAL OF RAILWAY PASSENGER SERVICES

The Minister of Transport has given his assent to the Board's proposal to discontinue all passenger train services between.

## BRISTOL (TEMPLE MEADS)
## and
## PORTISHEAD

and from the stations and halts at Ashton Gate, Clifton Bridge, Ham Green, Pill and Portishead, in the following terms:

Ministry of Transport,
St. Christopher House,
Southwark Street,
London S.E.1.
16th July, 1964.

RB. 3/4/0112

Sir,
    I am directed by the Minister of Transport to refer to the report of the Transport Users' Consultative Committee for the South Western Area upon objections and representations relating to the proposal to discontinue all railway passenger services between Bristol Temple Meads and Portishead, involving the discontinuance of all railway passenger services from the stations and halts at Ashton Gate, Clifton Bridge, Ham Green, Pill and Portishead. This proposed discontinuance is referred to in this letter as " the closure."

2.     The Minister has considered the report of the Consultative Committee and all other relevant factors. He accepts the view of the Committee that having regard to the bus services at present being provided such hardship as might arise from the closure would be alleviated by the provision of certain additional bus services. He has therefore decided to give his consent to the closure subject to the conditions mentioned below which, inter alia, require the provision of the additional bus services to which the Consultative Committee referred.

3.     Accordingly the Minister, in exercise of his powers under section 56 of the Transport Act 1962, hereby gives his consent to the closure subject to the following conditions:

(i)     The closure shall not take place unless and until the provision of the additional bus services set out in Part II of the Annex of this letter (hereinafter referred to as " the additional bus services ") has been authorised by road service licences granted under the Road Traffic Acts 1960-62 and until all necessary arrangements have been made to ensure that these services are available to the public immediately upon the closure taking place.

(II)     Whenever the Board become aware:
    (a) of any proposal for an alteration of any of the bus services at present being provided which are set out in Part I of the Annex hereto (whether they are being provided by the persons named in Part I or by any other person) or of the additional bus services set out in Part II of the Annex by withdrawing or substantially reducing the frequency of any such service, or
    (b) of any such alteration having been made.
the Board shall forthwith notify the Minister of any such proposal or alteration and give him all such information with respect thereto as he may reasonably require.

(III)     The Board shall take reasonable steps to keep themselves informed of any such proposal or alteration as is mentioned in the last foregoing condition.

(iv)     The foregoing conditions shall have effect until the Minister notifies the Board that they are no longer to apply or until they are varied under section 56(11) of the Transport Act 1962.

Notice published in July 1964 re the withdrawal of passenger traffic.

4.   I am directed by the Minister to make it clear that in imposing the conditions contained in paragraph 3 of this letter the Minister has been concerned only to discharge his functions under section 56 of the Transport Act 1962, and that these conditions have been framed in the light of the information before him for that purpose.   In particular it should be clearly appreciated that the additional bus services can only be provided and the existing bus services maintained in so far as their provision and maintenance is authorised by road service licences issued by the Traffic Commissioners under the Road Traffic Acts and that nothing in this letter affects the powers and duties of the Traffic Commissioners under those Acts in relation to the provision and maintenance of any of these services.   Furthermore in the event of any appeal to the Minister from any decision of the Traffic Commissioners the Minister will deal with that appeal in accordance with the provisions of the Road Traffic Acts and in the light of the information properly before him on the appeal.   Should the additional services eventually authorised by road service licences differ from those services required by the conditions contained in paragraph 3 of this letter, those conditions would be open to reconsideration and variation by the Minister.

I am, Sir,
Your obedient Servant,
C. R SCOTT-MALDEN
An Under Secretary,
Ministry of Transport.

The Secretary
British Railways Board,
222, Marylebone Road,
London N.W.1.

### ANNEX — PART I
Existing bus services provided under road service licences granted under the Road Traffic Acts 1960-62.
Services operated by the Bristol Omnibus Co. Ltd.

| Service No. | | |
|---|---|---|
| 85 | Bristol (Bus Station)—Pill—Portishead (via West Hill) |
| 85A | Bristol (Bus Station)—Pill—Portishead (via Nore Road) |
| 85B | Bristol (Bus Station)—Ham Green Hospital—Pill—Portishead (via West Hill) |
| 85C | Bristol (Bus Station)—Ham Green Hospital—Pill—Portishead (via Nore Road) |
| 125 | Bristol (Bus Station)—Falland—Portishead. |

### ANNEX — PART II
Additional bus services.

1. Services following the same route as either Service No. 85 or Service No. 85A referred to in Part I of this Annex between Bristol Bus Station and Portishead Railway Station as follows:
    Monday-Friday ex Portishead Railway Station—Bristol three additional services departing between 7.15 a.m. and 8.15 a.m.
    ex Bristol Bus Station—Portishead five additional services departing between 4.10 p.m. and 5.50 p.m.

2. Services between Bristol Temple Meads and Portishead Railway Station with stops at Pill and Ashton Gate as follows:
    Monday-Friday ex Portishead Railway Station—Bristol approximate starting time 7.30 a.m.
    ex Bristol Temple Meads approximate starting times 4.43 p.m. and 5.20 p.m.

NOTE: In this Annex the expression "approximate" in relation to a time specified for any service include any reasonable variation of the time so specified having regard to the class or classes of passengers likely to be carried by the service at the specified time.

The passenger train service will be withdrawn on and from
Monday, 7th September, 1964.

G. F. FIENNES,
General Manager.

Paddington Station.
July 1964.

had to bring in more supplies by rail while Albright &Wilson had its phosphate discharged at Avonmouth and brought to Portishead by rail. The railway ceased to enjoy this extra traffic when the docks were reopened on 1st March, 1959. The docks were busy enough for BR to keep a shunting engine there throughout the week. As on Saturdays and Sundays this engine dealt solely with the Central Electricity Generating Board traffic, the CEGB defrayed the cost of both engine and crew.

The 1960s saw passenger traffic generally tailing off, despite the introduction of diesel-multiple units (dmu). In March 1962 the passenger service was cut dramatically from 14 trains daily, to just six services run only at peak hours, while the five Sunday services were completely withdrawn. Then on 30th April, 1962 Portbury station was closed, while on 29th October, 1962 Ashton Gate Platform and Clifton Bridge became unstaffed.

The line still remained popular on Bank Holidays, for example on Whit Monday 1963 the 11.40 am from Bristol carried 400 passengers; the 1.15 pm from Portishead carried 200 passengers, the 2.30 pm from Bristol carried 180 passengers, and the 5.15 pm to Bristol 350 passengers (*see Appendix 6*). Although not advertised as such, the 5.15 was actually a through working from Portishead to Bournemouth, the same coaches throughout, but as Bath Green Park was a terminus, another locomotive worked it south of Bath.

Proposals were published for passenger traffic to be withdrawn on 3rd February, 1964. Traffic levels at the time were approximately 160 passengers daily Monday to Friday and 60-110 travelling on Saturdays. Annual revenue was £4,700 while costs were £16,100. In February 1964 Ernest Marples, Minister of Transport informed the House of Commons that he was delaying the closure of lines to certain holiday resorts in order to avoid uncertainty about holiday travel, so closure of the Portishead line was stayed until 7th September, 1964.

The line was competitive and should have been better supported. A day return by rail from Portishead to Bristol cost 3s. return compared with 4s. by bus and the train journey took only 28 minutes compared with 55 minutes by bus. However, the bus dropped passengers at The Centre, while the train took them to Temple Meads some distance from where most passengers were heading. But in the harsh winter of 1963 the A369 had been blocked by snow whereas the railway remained open.

Ironically the new Portishead station which had been only opened on 4th January, 1954 at a cost of £250,000 was converted to a garage and re-opened for this use on 26th August, 1966. Plans for its conversion had received strong local opposition as it was hoped that some sort of rail service could be reinstated.

# Chapter Eight

## The Freight-only Years

In June 1971 the freight side of the branch was looking rosy. St Anne's Board Mill at Brislington had hitherto been supplied with wood pulp via the Bristol City Docks from where it was taken up river by lighter to the factory, but with the forthcoming closure of the city docks to commercial traffic an alternative means had to be devised.

A five-acre site at Portishead was leased to St Anne's Board Mills and from this terminal, wood pulp was loaded direct to rail and moved in block trains to former carriage sidings at Marsh Ponds, east of Temple Meads and conveniently near the board mill. It was ideal because traffic carried in block trains was just the sort of traffic that Dr Beeching was seeking when he modernized the railways.

The traffic amounted to some 55,000 tons annually. Originally the fleet consisted of 52 specially-built wagons to carry the bales of wood pulp, but this eventually grew to 57. The bales were not roped, but either wire-whipped, or in later days, steel-banded. This traffic ceased in 1977. Coal traffic had stopped when the 'A' power station closed in March 1976 while the remaining 'B' station was oil-fired and its fuel arrived by sea. Other freight on the branch declined, cement being the only traffic in the final months, with the last train running to Portishead on 30th March, 1981 and leaving on 3rd April, 1981. Following that date a few occasional special Freightliner trains were worked, containers being road-hauled from the Royal Portbury Dock and placed on rail at Portishead, the line

Class '25' No. 25112 at Portishead beside timber wagons on 6th October, 1975.
*W. H. Harbor*

Portishead, view up from buffers to run-round loop, 21st March, 1987.   *Author*

Portishead, view to buffers, 21st March, 1987.   *Author*

being retained by BR as 'out of use', but it was hoped that traffic from the docks would increase and that it would appear in the proposed Avon Metro scheme. Steam-hauled and dmu passenger trains ran to Portishead in 1985 as part of the GWR 150 Celebrations.

Under a works order of 30th June, 1988 rocks at the west end of Clifton No. 1 tunnel were underpinned and rock fissures grouted to safeguard the tunnel portal. In 1990 a section of track was lifted to create a level crossing at Ashton Gate when the public footbridge became unsafe, though this rendered the branch unusable.

On 11th May, 1989 a Bill received Royal Assent to open a Metro line from Wapping Road to Portishead via Portbury at a cost of £28.5m. Although the re-opening of the line to passengers was deferred, the development of the Royal Portbury Dock meant that rail access was highly desirable so it was decided to re-open the line from Parson Street Junction to Pill and build a short branch to the dock. (For details *see page* 133.)

# Chapter Nine

## Description of the Line

Following the 1930s extension, most passenger trains to Portishead started from either Platform 1, 2 or 4 at Bristol Temple Meads, (118 miles 28 chains from Paddington). Leaving the station, a Portishead train travelled along the main West of England main line, to Bedminster (119 miles 22 chains) and tickets of passengers travelling to Temple Meads were collected here. The next station, Parson Street Platform, (120 miles 15 chains from Paddington), opened on 29th August, 1927 to serve a growing suburb. A 'Platform' was technically a 'Halt' which was manned for at least part of the time it was open.

Parson Street built in a cutting, had two timber platforms with a wooden shelter on the down side and one of corrugated-iron on the up. The porter-in-charge issued tickets from a small road-level office at the head of the steps leading from the up platform. Due to quadrupling, this

Aerial view of Temple Meads, 1935. Left to right: newly-built platforms with canopies; the 1870s curved train shed and Brunel's original terminus. *Author's collection*

A 4-6-0 passes through Bedminster with a down express.  *Author's collection*

Bedminster, view up, towards Bristol, after quadrupling.  *Author's collection*

# DESCRIPTION OF THE LINE 57

Bartlett's bridge blown up 26th April, 1931 in preparation for quadrupling.
*Author's collection*

The timber platforms and minimal shelters at Parson Street before quadrupling.
*Dr. C. L. Mowat*

Parson Street under construction. *Author's collection*

station was destined to have only a short time in its original condition. There were never any goods facilities.

Rebuilt to a four-road station and opened on 21st May, 1933 without the suffix 'Platform', it had two island platforms with covered waiting shelters. On Parson Street road bridge between steps down to the platforms, was the station offices in art deco style. A curiosity on the frontage was the use of 'GW' instead of the usual 'GWR'. Following a decline in station footfall, the offices and the canopy-type shelters were demolished in January 1971 and replaced with small waiting shelters. The outer down road, down relief line, was taken out of use 15th May, 1971.

At 120 miles 26 chains is Parson Street Junction, named Portishead Junction until a new signal box opened on 13th November, 1932 when the line was quadrupled to this location. Here the double track Portishead branch curves in a cutting, sharply away from the main line, falling at 1 in 97 easing to 1 in 76 almost to Ashton Junction signal box. Trains running on either the down main, or down relief could be signalled to the Portishead branch by Parson Street Junction signal box. To avoid a down train for the Portishead branch blocking the main line if waiting for an up train to clear the branch, a train-length of the branch from Portishead Junction was double track to enable it to on the branch. On 2nd September, 1883 the double track was extended to Clifton Bridge, the work being carried out by the Bristol firm of J. Durnford & Son, the overbridges had thoughtfully been built for double track. As freight traffic developed from Portishead, it was convenient to have a

DESCRIPTION OF THE LINE 59

Two ex-GWR diesel railcars, probably bound for Portishead, call at Parson Street *circa* 1954.
*M. J. Tozer*

Parson Street, view Up *circa* 1960. Notice the station offices on the bridge at the head of the steps. *Lens of Sutton*

View from Bedminster Down Road bridge, 21st August, 1980; Bristol West Depot is in the centre and the Portishead branch curves to the right.  *Author*

curve to offer a direct run to the West, so on 4th October, 1906 the West Curve opened, to a new Bristol West Loop South Jn signal box, about ½ mile west of Portishead Jn, thus forming triangular junction. Although normally only used by goods trains, it was cleared for use by passenger trains, Portishead – Weston-super-Mare Sunday school excursions being an example. West Loop North Junction signal box (120 miles 62 chains) was approximately on the site of the temporary Ashton Sidings East signal box opened 9th August, 1903 to 20th May, 1906 to control movements necessary for building the new curve. West Loop North Junction signal box closed 10th May, 1936 when the points were power-operated from Ashton Junction signal box situated a half mile beyond.

West Loop North ground frame (120 miles 52 chains), was brought into use on 25th January, 1938, to control Northville Building private siding, later Bristol Saw Mills & Transport Company, and May & Hassell Limited; the agreement was terminated 30th November, 1966. The West Curve of the triangle was closed 6th December, 1971, though much of one road remained until lifted in March 1980. The double track from Parson Street Jn to Ashton Jn remains as such.

Beyond West Loop North signal box Ashton Siding South ground frame (120 miles 69 chains) controlled the Imperial Tobacco Company Limited's private siding worked by an agreement of 27th July, 1918, the

The engine, with or without van, must be sent via Pilning Junction and carry "G" headmarks, and be so signalled on the block telegraph, in addition, the word "Banana" must be sent from box to box on the telephone or single needle to indicate that the engine is required for a Banana special.

The light engines or engines and vans, for the Banana specials must have precedence of all goods trains, and signalmen must advise the Signal Box in advance by telephone or otherwise if the engine is likely to be delayed.

(e) The Goods Agent is responsible for advising each Station of the departure of the Banana specials for Swindon and beyond as follows :—

"Banana special hence .. .. .. .. .. .. .. .."

He must also send a wire to Messrs. MORRIS "E" and HART (Paddington), Station Master (Reading), and Station Master (Didcot), giving the time of departure.

When such specials have traffic for the South Eastern and Chatham Railway via Reading, he must also add to the wire to Station Master, Reading, particulars of the traffic to be transferred at that Station.

Swindon must also wire Reading, giving time the Banana specials pass that Station.

(f) Signalmen must give every possible attention to the running of the trains, and communicate with the Signal Box in advance if they are likely to be delayed by other trains being ahead.

(g) The Locomotive Department will as far as possible arrange for passenger engines to work the Banana specials, and the light engines, running from Bristol to Avonmouth for such specials, must not under any circumstances be interfered with.

## Ashton Crossing—Placing Traffic in Giles' Siding.

When Trucks are shunted into Giles' Siding they must remain coupled to the Engine until they have been pushed back just inside the Catch Points where they must be detached for the Siding Owners' horse to take them towards the Works. Trucks must not be loosely shunted back into this Siding beyond the gates.

## Regulations for Blasting at the Netham Stone Quarries, situated between Clifton Bridge and Pill Stations on the Portishead Branch.

**1.** The Quarries to which these Regulations will apply are :—

No. 1.—Situated near the Portishead end of Tunnel, about one-third of a mile from the Netham Quarry Signal Box on the Clifton Bridge side of the Signal Box.

No. 2.—Situated about 400 yards from the Netham Quarry Signal Box on the Clifton Bridge side of the Signal Box.

No. 3.—Situated about 200 yards from the Netham Quarry Signal Box on the Portishead side of the Signal Box.

All being situated on the side of the Line opposite the River.

**2.** A Box is situated on the River side of the Line about one-third of a mile on the Portishead side of the Tunnel, from which Fixed Signals, the normal position of which is "All Right," will be worked. There is also a Bell communication between Netham Quarries and Clifton Bridge, and a Keyless Disc fixed in the Netham Quarries Box to denote to the Signalman there the state of the Line on each side of him.

**3.** Blasting operations must only be carried on under the sanction of the Signalman stationed there.

**4.** A Blasting Disc, worded "Blasting Disc for Netham Quarries," is in use, and except when in possession of the Foreman in charge for the time being of the blasting as his authority to blast, it must always be kept in the personal charge of the Signalman in the Signal Box. The Foreman in charge of the blasting is the sole person authorised to receive the Disc from the Signalman, and under no circumstances must shots be fired at either Quarry which might cause rock to fall on to or across the Line, unless the Foreman in charge is in possession of the Disc, and he must keep it in his possession until the whole of the shots have been fired, and any obstruction upon the Line has been removed.

**5.** Small shots for blasting on the ground or from the high rock, which are not likely to throw stones on to or across the Line, may be fired without the Blasting Disc, provided such shots are not fired within 15 minutes of a Passenger Train being due to leave Clifton Bridge or Pill, nor until such Train has passed the Quarry.

**6.** The Signalman at Netham Quarries must not allow the Blasting Disc to go out of his possession unless the blasting can be carried out and the Line cleared at least 30 minutes before a Train is due to leave the Station on either side.

**7.** Before the Signalman at Netham Quaries hands the Disc to the Foreman in charge of the blasting, he must send to Clifton Bridge the Code "Blasting required to be carried out" (4 beats on the Bell, as per Bell Code). On receipt of this Signal, the Signalman at Clifton Bridge must send to Pill the Code "Can Staff be withdrawn for blasting purposes?" (4 beats on the Bell, as per Bell Code), and when Clifton Bridge is able to draw the Staff, the Signalman there must acknowledge the Bell Signal to Netham Quarries, when the Blasting Disc may be handed to the Foreman in charge of the blasting, after placing his fixed Signals at "Danger."

**8.** The withdrawal of the Staff at Clifton Bridge will place the indicator at Netham Quarries to "Down Train on Line," in which position it will remain while the Staff remains out of the Pillar.

9. After the blasting has been completed, and the Signalman has satisfied himself that the Line is clear, he must send to Clifton Bridge the Bell Signal, " Blasting complete, Line clear " (3 beats on the Bell, as per Bell Code), which must be acknowledged by Clifton Bridge, when the Signals may be placed at their normal position. On receipt of the Signal, " Blasting complete, Line clear," from Netham Quarries, the Signalman at Clifton Bridge will send the " Cancel Last Signal Sent " Signal to Pill and replace the Train Staff in the Pillar, after which the usual working may be resumed.

10. If on receipt of the Bell Signal from Netham Quarries permission cannot be given to blast the Signalman at Clifton Bridge must send the " Obstruction " Signal (6 beats on the Bell, as per Bell Code).

11. Should any accident occur at the Quarries by which the Line becomes blocked the Signalman at Netham Quarries Box must immediately send the " Obstruction " Signal (6 beats on the Bell) to Clifton Bridge and place his fixed Signals to " Danger " in which position they must remain until the ' Obstruction " is removed.

12. On receipt of the " Obstruction " Signal from Netham Quarries, the Signalman at Clifton Bridge must immediately send the same Signal to Pill, and place the whole of his fixed Signals at " Danger " to stop any Train going towards Netham Quarries.

13. When the obstruction has been removed the Signalman at Netham Quarries must send to Clifton Bridge the " Obstruction Removed, Line Clear " Signal (as per Bell Code), which Signal must be repeated to Pill, and then Trains may be allowed to come forward.

14. The Signalman at Netham Quarries must, in the event of the line becoming blocked, send an immediate advice by telegraph to the Station Master at Clifton Bridge, and also to Divisional Superintendent, Bristol.

15. If the men engaged at the Quarries carry out any blasting operations without being in possession of the Blasting Disc, except as provided in Regulation 5, or if they in any way depart from these instructions or do anything to interfere with the safety of the Line, or if they neglect to obey the orders given, the Signalman must report the circumstances to the Station Master at Clifton Bridge, who must immediately report to the Divisional Superintendent, Bristol.

16. No blasting must be carried on before daylight or after dark or during a fog.

17. Should the Telegraph Single Needle, Bell, or Electric Staff Instruments be out of order, or the fixed Signals in any way defective, so as to prevent the necessary protection to Trains being given, all blasting operations must be stopped until such have been restored.

18. The Signalman will be required to be on duty at such hours as the men are appointed to work at the Quarries, which will vary at certain seasons.

19. **BELL CODE.**

|  | No. of Beats. | How to be given. |
|---|---|---|
| 1. To call attention | 1 |  |
| 2. Blasting required to be carried out | 4 | Consecutively. |
| 3. *Can Staff be withdrawn for Blasting purposes ? | 4 | 2 pause 2. |
| 4. Blasting complete, Line Clear | 3 | 2 pause 1. |
| 5. Cancel Signal last sent | 8 | 3 pause 5. |
| 6. Obstruction Signal | 6 | Consecutively |
| 7. Obstruction removed, Line Clear | 3 | 2 pause 1. |

These Signals will be used by Clifton Bridge, Pill, and Netham Quarries, except that marked thus * which will be used only by Clifton Bridge and Pill. Nos. 2 and 4 Codes must not be used by Pill.

### Corporation Quarries and Hickory's Quarry adjoining Portishead Line.

This Quarry is situated on the Down Side of the Line about 350 yards on the Portishead side of the Netham Quarries Signal Box. The Blasting operations are carried on a considerable distance from the line but to ensure absolute safety for the passage of Trains and this Company's Staff working on the line in the vicinity of the Quarry the following regulations must be strictly carried out.

Shots must only be fired during the day and immediately after the passing of Passenger Trains in accordance with the Company's Time Table, and before such Trains are due to leave Clifton Bridge Station for Pill, or Pill for Clifton Bridge Station, all arrangements for firing must be completed, and a man from the Quarry must be sent in each direction at least 100 yards to warn any person on the Line that shots are about to be fired, and they must wait there until the Passenger Train has passed and until all the holes charged have been fired. A man must also be stationed opposite the Quarry to give the Signal for the firing of the shots after ascertaining from the men who have gone out in either direction that the Line is clear of workmen and that the Train has passed. No blasting operations must under any circumstances be carried out if a dense fog, or thick snow storms exist, and only during broad daylight.

## 148

## PORTISHEAD.
### Instructions for working the Gas Company's Siding near Portishead Station.

1. The Siding is provided with throw-off points, which are coupled to the points leading to the Main Line, and is worked from a Ground Frame which is secured with Annett's Patent Lock, the Key being attached to the Train Staff, and no Engine Driver must start for the Siding unless he is in possession of a Staff.

2. When the Siding is to be used, instructions will be given by the Station Master to the Signalman at Portishead to withdraw a Train Staff for the purpose by sending the "Release Staff for Shunting" Signal to Pill, as per Electric Train Staff Standard Instructions, Clause 13, shown in the Appendix to the Book of Rules and Regulations. The Staff must be handed to the Driver to proceed to the Siding and a competent Guard and Porter must ride in the last Vehicle from Portishead, and the first on returning from the Siding.

3. The Guard will be in charge of the working, and must see that everything is in order, and the Points properly set for the Main Line before returning to Portishead.

4. On the return of the Train from the Siding, the Train Staff must be dealt with in accordance with the Instructions referred to in Clause 2.

### Working of Bristol Corporation Light Engine over the Main Line between Dock Sidings and Timber Wharf Sidings.

By an Agreement dated March 11th, 1903, the Bristol Corporation Light Engine may be allowed to work over the Main Line between the Docks Sidings at Portishead Station, and the Junction with the New Timber Depot, situated about 500 yards on the Bristol side of the Station. Each time the Engine is required to run on to the Main Line, an application will be made by the Corporation Staff to the Signalman at the Station Signal Box, and subject to the working of this Company's traffic not being interfered with, the Engine may be allowed to proceed without delay, in accordance with the following Regulations:—

(1) Under no circumstances must the Engine be allowed to pass on to the Main Line at either end without a competent Porter or other Company's Servant riding on same, and he must remain with the Engine during the whole time it is on the Main Line.

(2) Before the Engine is allowed to enter upon the Main Line at either end, the Porter or other Company's Servant in charge, must obtain a Staff for working between Portishead and Pill, and he must keep it in his possession until the Engine has been shunted clear of the Main Line, when he must return it immediately to the Signal Box.

(3) The Porter or other person, who accompanies the Engine to the Timber Wharf Junction is responsible for working the points, and for seeing that the Driver exercises care in passing to and from the Main Line.

(4) The speed of the Engine while running over the Main Line must not exceed 8 miles per hour at any point, and the Porter or other person who accompanies the Engine, must see that this speed is strictly observed.

Immediately a Staff has been handed to the Porter for the Engine to run from the Timber Wharf Junction to the Station, the Signalman must set the points leading from the Main Line to the Dock Sidings at the Station for those Sidings, and keep them in that position until the Engine has arrived, and the Staff has been returned to him.

The Signalman must record in the Train Register, the time application is made for the Engine to run, the time of withdrawal of the Staff for same, and when it is returned to him.

### Portishead.

Sidings are provided for the exchange of trucks between the Weston, Clevedon and Portishead Light Railway and the Great Western Railway, but no other trucks or traffic than that for exchange between the two railways named must be dealt with at these exchange sidings.

The key of the points leading to and from the sidings will be kept in Portishead Signal Box, and the points must be locked so as to prevent the use of the sidings when the key is in the Box.

Under no circumstances may the engines of the Bristol Docks Committee work into or out of the exchange sidings, nor may the engines of the Light Railway Co. work to and from the Corporation Sidings, nor beyond the exchange sidings referred to.

The Line between the Station and the Pier at Portishead is not to be used for Passenger traffic.

### Loads of Goods Trains on Portishead Branch.

The loads of the Goods Trains on the Portishead Branch must not exceed 28 wagons.

This applies to the loading of Trains between Clifton Bridge and Portishead only, and must be rigidly observed over that portion of the Line during the time the Branch is open for Passenger Trains. In cases, however, where it is necessary to run Goods Trains after the Passenger Train service has ceased for the day Goods Trains (Up or Down) may be made up to the number of wagons the Engines can take, provided the same are not appointed to cross Trains running in the opposite direction, at Pill.

## BRISLINGTON.
### Ground Frame.

Ground Frames are fixed on up side of Line at Brislington (one at each end of Station) for working the points leading to and from the Sidings.

The Frames are locked by key on Electric Train Staff, and the Station Master or Porter will be responsible for working same.

The points at the Bristol end of Sidings may only be used by Trains coming from the direction of Radstock.

## NETHAM, CORPORATION AND HICKORY'S STONE QUARRIES BETWEEN CLIFTON BRIDGE AND PILL. INSTRUCTIONS IN CONNECTION WITH BLASTING OPERATIONS FROM MAY 5th, 1924.—Addition to Page 147.

With reference to the instructions on pages 146 and 147 of No. 4 Section Appendix, dated January 1st, 1922, the three Netham Stone Quarries have been closed and blasting has ceased. The instructions, therefore, are cancelled.

The Corporation Quarry next to these in the direction of Pill and that known as Hickory's Quarry again next in the direction of Pill, are now being worked by Messrs. William Cowlin & Son under the following arrangements, which supersede those shewn at the foot of page 147 of the Appendix referred to.

A telephone in a hut has been fixed by Messrs. Cowlin & Son on the Pill side of 123¼ mile post at a point opposite Hickory's Quarry, where there is a pathway leading up to the G.W.R. Portishead Single Line, and whenever blasting operations are required to be carried out at the Corporation Quarry or Hickory's Quarry, a properly appointed servant of Messrs. Cowlin & Son will telephone from this hut to the Signalman at Clifton Bridge, with whom it forms means of communication, and ask permission to blast. The Clifton Bridge Signalman, before giving permission for the blasting operation to be performed, must (providing that no train is due to pass the Quarry within 30 minutes of the time the request is made), send the bell signal—2 pause 2 (" Can staff be drawn for blasting ") ? to Pill and withdraw the Electric Train Staff for the Clifton Bridge—Pill Section. After such telephone permission has been given, the Signalman at Clifton Bridge must not replace the Electric Train Staff in the instrument until Messrs. Cowlin & Son's watchman has informed him by telephone that the blasting has been completed and that the line is clear, when the Electric Train Staff must be restored and the bell signal—2 pause 1 (" Blasting complete and staff restored ") sent to Pill. During the time of such blasting Messrs. Cowlin & Son will provide a watchman to be posted on the railway line near the hut to see that no obstruction takes place and to inspect the line after each occasion of blasting.

The Signalman at Clifton Bridge must record in his Train Register Book the times and particulars of all messages received from and sent to Messrs. Cowlin's man.

Blasting will not take place after dark nor during fog or falling snow, and the most modern and up to date methods of conducting the blasting will be adopted by Messrs. Cowlin & Son.

Branch information in the Working Timetable Appendix & Supplement March 1910.

BR Standard Class '5' 4-6-0 No. 73166, apparently in not very good health, at Ashton Junction 22nd February, 1965.  *Paul Strong*

sidings came into use on 7th November, 1918. The agreement transferred to Ashton Saw Mills Limited on 12th December, 1919 and to Ashton Containers Limited on 23rd July, 1937. Ashton Containers Limited traffic ceased on 13th May, 1965.

At the level crossing over Ashton Vale Road was Ashton Crossing signal box, the centre of a busy complex: to the west was a siding to Ashton Vale Brick Works, later used by the coal mining department of the Ashton Vale Iron Company Limited. It was a beneficial arrangement allowing the coal to be dispatched directly from the pit instead of having to be carted along Ashton Road to the Bristol & Exeter Railway. From 8th June, 1942 until 10th September, 1946 the former siding to the mine was used by the Ministry of Supply, then by Messrs F. W. Toogood 16th March, 1948, Rudders & Payne 1955 and finally 26th March, 1954 by Strachen & Henshaw Limited.

To the east of the Portishead branch were sidings to Giles & Son's nail works and Ashton Vale Iron Company's rolling mills. The nail works siding agreement was terminated 4th December, 1935 and the rolling mills siding 9th March, 1931.

On 9th August, 1903 Ashton Sidings West signal box opened as a temporary expedient to control access to the Bristol Harbour Railway under construction (*see Chapter 15*). This box closed on 20th May, 1906 when it, and Ashton Crossing box were replaced by Ashton Junction signal box (121 miles 18 chains), which opened 20th May, 1906, on the Portishead side of the crossing. The box was built on two timber beams over a stream and was always damp. The Bristol Harbour lines opened 4th October, 1906. The gates over Ashton Vale Road, were replaced by lifting barriers on 12th March, 1978.

At Ashton Junction the Bristol Harbour Railway and Canon's Marsh branch curved away from the Portishead line and immediately beyond was Ashton Gate Platform, (121 miles 30 chains).

With the promotion of Bristol City Football Club into the Football League's First Division, it was anticipated that the ground would experience more spectators arriving from various parts of the kingdom, so convenient railway facilities needed to be provided. The timber Ashton Gate Platform opened on 15th September, 1906 to football excursions and the public on 1st October, 1906. Had the First World War not put a stop to the Bristol International Exhibition in 1914, Ashton Gate would have been developed as the Exhibition station. It opened partly on 13th June in order to receive some return on the money spent on it. The Dominions building contained fine pictures and Bristol Castle naval and military treasures, while mural canvasses measuring 50 feet

A Class '08' diesel-electric shunter passes Ashton Junction signal box and level crossing 18th April, 1970. *D. Payne*

by 30 feet were displayed in the open. A menagerie and 'coastal railway' catered for more youthful tastes. On 29th June, 1914 the full exhibition including the International Pavilion opened on Ashton Meadows. Thousands of visitors arrived, many by train, but it suddenly closed on 15th August, 1914 when the ground was taken over by the War Office as barracks for army recruits.

As a wartime economy measure Ashton Gate Platform closed on 1st November, 1917, but rebuilt, it re-opened 23rd May, 1926, while in August 1928 it became simply Ashton Gate in the timetable though the name board still announced 'Platform'. The many exits and spacious footbridge from the 400 foot long platforms, helped to disperse large crowds rapidly. It was staffed from 1st January, 1929, the booking office being at road level; the staff was withdrawn 29th October, 1962.

In the 1930s, when heavy football traffic was expected, to ease crowding an up train could be started from the down platform under supervision of the inspector in charge and with a groundman at the crossover points to clip them for the passage from the down platform to the up road.

Following closure of Ashton Gate on 7th September, 1964, due to traffic difficulties which would be experienced from too many supporters arriving by road, it was reopened for football specials. The down platform which had been shortened by road improvements, was

Handbill issued September 1957 showing fares to Ashton Gate.

Despite the caption, the platform sign reads 'Ashton Gate Platform'. The railmotor's headboard reads 'Bath'. A 0-6-0ST stands on the Bristol Harbour and Canon's Marsh lines (*right*). This view was probably taken shortly after the platform opened in 1906.  *M. J. Tozer*

Class '3' 2-6-2T No. 82037 working the 2.15 pm Portishead to Temple Meads, crosses the 2.30 dmu Temple Meads to Portishead, 7th June, 1960.  *R. E. Toop*

DESCRIPTION OF THE LINE 69

'8750' class 0-6-0PT No. 4603 at Ashton Gate heading a train from Canon's Marsh, 15th June, 1960.  *E. T. Gill*

21-ton hopper wagons stored on up line at Ashton Gate, 21st August, 1980.  *Author*

# THE BRISTOL TO PORTISHEAD BRANCH

Ashton Gate Platform view towards Ashton Junction, 6th May, 1987. Only the right-hand platform (former down line) had been in use following reopening in 1970.  *Author*

Class '2' 2-6-2 No. 41203 approaches Ashton Gate with an up train 17th June, 1960.  *E. T. Gill*

cleaned, spread with chippings and connected to the road by a new footpath. The first train after re-opening, a 3-car dmu, arrived from Birmingham on 29th September, 1970, but in 1977 Parson Street station took over the football specials and Ashton Gate again fell out of use. From 12th-19th May, 1984 when Billy Graham spoke at the Ashton Gate Stadium during Mission England, the platform was again re-opened to assist those arriving by train.

From Ashton Gate Platform the line rises on a ruling gradient of 1 in 100 to a summit half a mile beyond Oak Wood signal box. Immediately Bristol side of Ashton Gate Platform, the line was singled 2nd May, 1965 (single line junction at 121 miles 25 chains), officially closed 3rd August, 1981 and taken out of use, but retained *in situ* 5th December, 1983. Ashton Junction ground frame (121 miles 37 chains) gave access to three carriage sidings brought into use on 2nd April, 1944, two more being added on 3rd October, 1958. Two sidings generally held suburban stock, the others holding corridor stock such as was used for excursions. The local ganger's cottage was nearby.

In 1940 the Ministry of Works & Buildings spent £18,000 on land between Ashton Gate and Clifton Bridge on which to build a new station to entrain workmen for the Bristol Aeroplane Company's underground factory at Spring Quarry, Corsham, Wiltshire, and sidings for stabling these carriages. It was then observed that the line between Bristol and Bath was so intensively used that there was no paths for such trains, so a fleet of 120 buses was used instead.

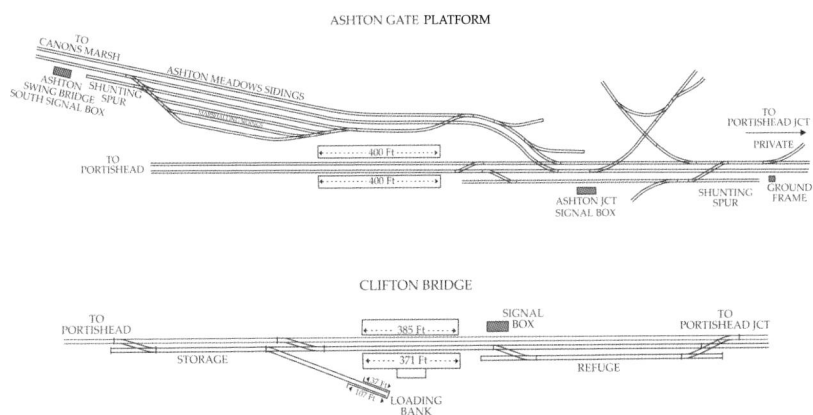

Plans of Ashton Gate Platform and Ashton Meadows Sidings; Clifton Bridge station.

Clifton Bridge 1866: the impermanent way and contractor's wagons. The New Inn purchased by. the railway in 1862 became the office for the resident engineers.
*Author's collection*

A down broad gauge passenger train hauled by a 2-2-2WT calls at Clifton Bridge *circa* 1867. The contractor's plant stands on the left for auction. *Author's collection*

## DESCRIPTION OF THE LINE

Clifton Bridge station (121 miles 63 chains) opened 18th April, 1867 and closed to passengers 7th September, 1964 and goods 5th July, 1965. It was unstaffed from 29th October, 1962. As Clifton Down station did not open until 1874, Clifton Bridge was the nearest station to the prosperous Clifton suburb. To avoid confusion with Clifton Down station it was renamed Rownham in March 1891, but reverted to its original name in 1910. The station offices were on the down side: the offices on the ground floor with the station master's private accommodation above. The signal box (dating from 1879) was a Furness Railway design, due to the engineers being of Furness origin.

Following gauge conversion 24th-27th January, 1880 on 15th September, 1880 a crossing loop and up platform were brought into use. Both platforms were covered for part of their length by 'half-barrel' type canopies. The loading dock had the capacity for two four-wheeled wagons. The footbridge at the up end of the station doubled as a link between the platforms and also to carry a public footpath across the railway. The original signal box on the up platform closed 25th August, 1907 being replaced by one just beyond the ramp of the up platform. The box contained three token instruments to Oakwood, Pill and Portishead respectively, as one, or both, of the intermediate boxes could be switched out, this economic arrangement dating from 1932. Oakwood normally

A '481' class 2-4-0T at Clifton Bridge with an Up train in the 1890s. Note the suspension bridge in the background.   *Author's collection*

The timber footbridge immediately south of the station carrying a public path. The signal has a slotted post. This view was taken before a new signal box was opened between the footbridge and station.
*M. J. Tozer*

The penultimate passenger train from Portishead calls at Clifton Bridge, 5th September, 1964.
*E. T. Gill*

DESCRIPTION OF THE LINE 75

Clifton Bridge view up *circa* 1965. Beyond are the carriage sidings, while on the far right is a 10 mph speed restriction sign. *Lens of Sutton*

Clifton Bridge station following the removal of platform canopies. *Author's collection*

The site of Clifton Bridge station, view up, 21st March, 1987.   *Author*

A Portishead to Temple Meads dmu approaches Clifton Bridge on 30th April, 1959. Six carriages are stabled on the left. A steam tug can be seen on the right.   *E. T. Gill*

DESCRIPTION OF THE LINE 77

The ferry linking Hotwells with Clifton Bridge station. Clifton Bridge No. 1 tunnel can be seen centre-left. *Author's collection*

only worked the morning shift, Pill closed overnight, only Clifton Bridge and Portishead being worked for 24 hours.

Although the station seems in a picturesque, rural situation at the southern end of the Avon Gorge, in addition to being served by horse buses it could be reached via Rownham Ferry to and from Hotwells. The chain ferry carried employees of Portishead power station and the docks to Clifton Bridge station, while in addition to those from Portishead who worked in the centre of Bristol it could be quicker to get off the train at Clifton Bridge, use the ferry and then the electric tram, taking a total of less than 40 minutes from Portishead, whereas if they went by rail to Temple Meads they would have been delayed by ticket inspection at Bedminster and then by their train having to wait for a vacant platform at Temple Meads. A favourite outing for Bristolians was crossing Rownham Ferry to Clifton Bridge and either having a scenic rail journey through the gorge, or walking down the riverside footpath.

Clifton Bridge station was particularly important on several occasions. In 1878 when the Prince of Wales visited the Royal Agricultural Society's Show held on Clifton Downs, he travelled by road from Temple Meads to the show ground and then afterwards crossed the Suspension Bridge to Clifton Bridge station where a special train took him onwards.

In the First World War the station became a depot for all mules arriving at the local docks before they were taken by rail to train at such places as Shirehampton or Salisbury Plain.

The importance of Clifton Bridge is shown by the fact that in the 1930s an average of eight staff were based at the station. Cash taken by the booking office would be placed in a leather bag, then slipped inside a travelling safe which had a non-returnable lid and was pad-locked and chained to the guard's brake of an up train to Temple Meads, wages would arrive in a down train the same way. The other branch stations used the identical procedure.

The branch grew in importance during the Second World War as in addition to wartime traffic using Portishead Docks, Bristol residents who had been evacuated to places along the line, used the branch to travel to work in Bristol each morning. When air raids were threatened, some sheltered in carriages stored in Ashton Meadows sidings – though had a bomb fallen nearby and shattered the windows, they could have been badly lacerated by glass.

Members of the Women's Land Army loaded timber from the nearby woods on to wagons at the station. Traffic from the Ashton Containers factory was dealt with by Clifton Bridge, but accountancy was handled at Temple Meads.

The down platform was 371 feet long and the up 385 feet. The mileage siding held five wagons and at its end was a loading bank. A Down refuge siding at the Up end of the station held 25 wagons.

As the signal box was in the centre of the up platform, to ease the task of the signalman and fireman, the fireman of a down train which

BR Standard Class '3' 2-6-2T No. 82033 approaches Clifton Bridge with an up train from Portishead 4th July, 1959. No. 1 Tunnel is on the right.  *E. T. Gill*

crossed an up train at Clifton Bridge obtained the single line token from a platform token pillar fixed in a hut at the Portishead end of the down platform.

Following the station's closure, in 1971 it became the headquarters of the Avon & Somerset Mounted Police & Dog Section until moved from the site in August 2016.

The line climbs and just north of Clifton Bridge station enters the single bore 59-yard long Clifton Bridge No. 1 tunnel which passes under one of the bridge abutments. It is on a gradient of 1 in 100 rising towards Portishead and set on a 20-chain radius curve. It is dry, brick-lined and due to having being built to double track, broad gauge dimensions, though only a single track was laid, has a generous bore. During air raids in the Second World War, some Bristolians sheltered there.

One sub-species of the whitebeam tree has only 23 examples in the world, 21 of these grow on railway property in the Avon Gorge which the branch follows for three spectacular miles.

The *Bristol Times & Mirror* of 16th September, 1863 was critical of the railway's invasion of the Nightingale Valley saying: 'The age of the romantic and picturesque is certainly gone when stokers take possession of the sketcher's seat and you can have a return ticket for the meadows of the Templers from the valley of the Nightingales.' The railway, sensitive to the environment constructed the arch in Nightingale Valley with rustic masonry and covered it with ivy and appropriate plants.

A Portishead to Temple Meads dmu leaves the 59-yard long Clifton Bridge No. 1 tunnel 4th June, 1960. The 'cat's whisker' was the original warning sign on the front of the dmus and enabled men working on the track to estimate better the approaching speed.

*R. E. Toop*

In due course the attractive scenery was made available to a wider public when the picturesquely-named Nightingale Valley Halt (122 miles 37 chains) opened on 9th July, 1928. The *Bristol Times & Mirror* reported that 'it offers the opportunity of travelling, with the minimum of inconvenience and expense, to the centre of a scene of natural beauty unsurpassed anywhere in such close proximity to a great city'. It was only open during the summer months and due to lack of use, closed permanently on 12th September, 1932. Measuring 400 feet by 8 feet it was cheaply constructed with a wall of sleepers on edge forming a platform wall, while set every nine feet was an upright sleeper driven into the ground and tied back by a wire rope to a wooden peg. The structure was then backfilled with ashes to form the platform surface. A corrugated-iron shelter provided for the inclement weather likely to be experienced during a British summer. The estimate of £380 was in the event exceeded by £16 16s. 6d. Nightingale Valley was unstaffed and supervised by the Clifton Bridge station master. Takings for the first three months were:

| Month | Number of tickets | Income |
|---|---|---|
| July | 325 | £9 |
| August | 414 | £10 |
| September* | 195 | £5 |

*to 23rd when closed for winter

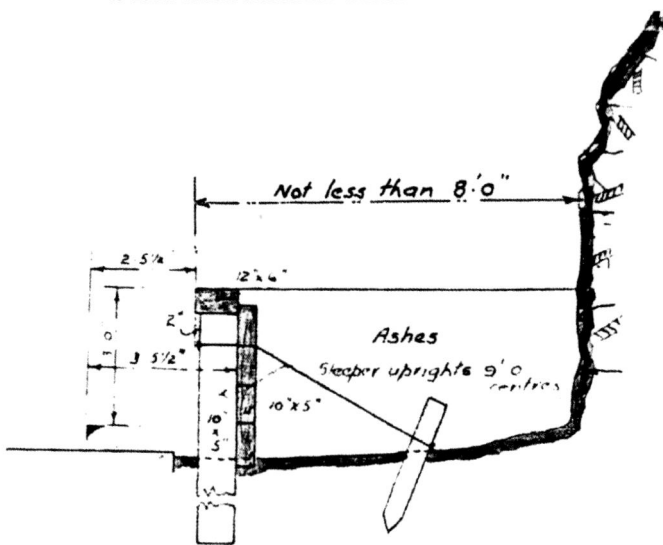

The section plan for Nightingale Valley Halt: a wall of sleepers on edge and every 9 feet a vertical sleeper is tied to a wooden peg. The site is then back-filled with ash.

# DESCRIPTION OF THE LINE 81

The 'Donkey Slider' at the mouth of No. 2 tunnel; although the rock was smooth, it still wore out trousers!
*Author's collection*

The beautiful area in which the halt was set was once in grave danger of having hundreds of small dwellings being erected. The Leigh Woods Land Company was then established to prevent the woods and valley from being exploited. In 1900 Sir George Wills of tobacco fame, purchased the property and vested it in trustees 'for the reasonable enjoyment of the public'.

Clifton Bridge No. 2 tunnel, 232 yards in length, on a curve of 20 chains radius is on a gradient of 1 in 300 rising towards Portishead. It is a wet tunnel lined only in the centre; at its further end is a smooth, sloping slab of rock known as the 'Donkey Slider'. The line then crossed a bridge over the Netham Chemical Company's tramway from Greenland Quarry to a jetty on the Avon. On the other side of the river the Bristol Port Railway & Pier line swings into view and the two railways run parallel for about a mile.

At 123 miles 5 chains Netham Quarries signal box opened in 1877 to serve quarry sidings and give warning of blasting, and long disused until its probable closing date of 9th January, 1922. On 29th September, 1876 an agreement had been signed between the railway and Thomas Wethered & Worsley regarding the working of the stone quarries. They were at liberty to blast rock only after the last train and before 3.00 am. At the far end of the siding the line crossed another of the Netham Chemical Company's tramways, this one from Pheasant, or Whitestone, Quarry to a riverside jetty.

Further on the branch crossed Bristol Corporation's tramway from Abbot's Leigh quarry to a jetty and Mr Hickery's tramway from Leigh Valley Quarry which ran to another jetty.

At 123 miles 77 chains the dry, 88 yard long Sandstone tunnel is negotiated on a gradient of 1 in 1056 rising to Portishead. The side walls

Oak Wood signal box view north 10th April, 1953. *Dr A. J. G. Dickens*

A down train hauled by a '8750' class 0-6-0PT passes Oak Wood signal box 3rd March, 1962. The down loop had been taken out of use on 25th September, 1960, when the signal box closed. *Hugh Ballantyne*

of masonry piers carried brick relieving arches, the panels within these arches being of natural rock. As the relieving arches were in poor condition they were rebuilt in 1934 and by 1969 the tunnel was brick-lined throughout. Beyond the tunnel the line crosses a four-arch viaduct. At 124 miles 62 chains was the passing loop at Oak Wood signal box opened 14th May, 1929 and closed 25th September, 1960 when the loop was taken out of use. The box, when opened, had electric token working to Clifton Bridge and Pill and was provided with a block switch. When switched out the long section Clifton Bridge-Pill continued to be worked by electric staff. The box had no piped water, so this commodity was brought by train. In the 1940s the 8.12 am down diesel railcar picked up the empty cans, filled them at Portishead and dropped the full cans on its return. During the hours the box was closed, trains used the up loop which gave a straight run. Just beyond this loop is the longest cutting – 830 yds in length and up to 50 ft deep. It was originally intended to be a tunnel but it was then discovered that the work could be more expeditiously and cheaply carried out by substituting a cutting. Where it is crossed by a three-arch overbridge the line almost immediately falls from its summit on a gradient of 1 in 98/92 to Ham Green Halt. Beyond the cutting is Chapel Pill viaduct, its five piers set on concrete foundations.

The timber-built, single platform unstaffed, unsheltered Ham Green Halt (125 miles 26 chains) was authorized in October 1926 at a cost of £220 and opened on 22nd December, 1926 – not 23rd December as has been quoted elsewhere. All trains called and offered a Sunday, as well as a

A down train of 7 coaches headed by a '45XX' class 2-6-2T leaves Ham Green Halt. A goodly crowd have stepped from the train. *Author's collection*

weekday service. As most passengers using it were expected to have return tickets, it was unstaffed. As the platform was so short, to expedite working, passengers were encouraged to use just one coach. A pathway served the hospital entrance (*see below*). The original nameboard simply announced 'Ham Green'. On 27th October, 1927 a stone-built extension of 300 feet was authorized together with an additional shelter and extra lighting, the extra work completed by August 1928, totalling £443. In 1939 electric lighting cost a further £105. Opposite the halt was a lily pond. The unstaffed halt came under the supervision of the Pill station master who arranged for the trimming and lighting of lamps, the guard of the last train being responsible for extinguishing them. Fares were collected at other stations – at Bedminster for stations beyond – but the guard collected Ham Green fares for passengers from there who were to alight at Nightingale Valley Halt.

In addition to serving Ham Green, it also served Ham Green Isolation Hospital. Due to numbers visiting the isolation hospital, the platform unusually had two platform shelters. During the Second World War part of Ham Green Hospital was used by United States' forces and this brought additional traffic to the halt. Opposite the platform was a lily pond known to railwaymen as 'Cripple Creek'. Beyond the halt the line rises at 1 in 162 through the 665-yard long Pill tunnel with an elliptical-shaped mouth. The tunnel on a 1 in 100 gradient rising towards Portishead is on a 20 chains radius curve. Brick-lined, half of it in engineering brick, a considerable amount of water percolates through the roof. Smoke cleared quickly, partly due to the 7 feet 6 inches diameter ventilating shaft 205 yards in from the Bristol end. As some of the hospital wards were built above the tunnel, patients could hear the rumble of trains passing below. During an air raid it was not unknown for a goods train to take shelter in the tunnel.

Originally Pill tunnel was to be only a cutting, but then the owners of Ham Green House insisted that the line be in a tunnel so that they could not see or hear the trains. In May 1985 clearance in the tunnel was found to be particularly tight on the Up side.

Approaching Pill station (126 miles 12 chains) is the 92 yard-long, 6-arch, brick-built Pill viaduct 43 ft above the village. One pier caused special difficulty as the land was boggy and required a concrete foundation 30 feet deep. Part of the passing loop crosses the viaduct. In addition to two centre check rails over the viaduct to prevent a derailed train plunging off, east of the viaduct two check rails are placed outside the running rails to divert a derailed vehicle to a cutting slope.

On 18th May, 1933 a block switch was installed in Pill signal box. As Oakwood, Pill and Portbury Shipyard could all then switch out if

# DESCRIPTION OF THE LINE

Ham Green Halt view down *circa* 1960. Notice the two corrugated-iron pagoda waiting shelters and the access path winding to the top of the cutting. Passengers using the seat by the running-in board would have to be careful not to strike their heads when rising. Unusually the second word of the name is in smaller letters. *Lens of Sutton*

View from Ham Green Halt to the 665-yard long Pill tunnel. The access footpath on the far left could do with some attention. A platelayers' hut is on the right, its window boarded to prevent vandalism. *M. J. Tozer*

View from Pill tunnel towards the viaduct; notice the abandoned hut. 29th June, 2001.
*Author*

required, the long section Clifton Bridge to Portishead was worked by electric staff.

Pill opened on 18th April, 1867, closed to goods on 10th June, 1963 and to passengers on 7th September, 1964; the signal box on the down platform closed on 14th April, 1964. It was sited conveniently in the centre of the village, and as the platforms were in a cutting, the main station building was at road level with brick shelters on both platforms. The platforms were extended by 100 feet to 303 feet in January 1880 and on 7th March, 1912 to 404 feet. The passing loop was extended at both ends on 7th March, 1912 making it 1,010 feet in length and at the same time a looped goods siding was laid at the up end of the station. In 1956 Pill had a 1½ ton hand crane in the goods shed. The mileage siding had a capacity of 20 wagons.

Up passenger trains could be signalled into either platform, a useful facility if an up train was not crossing another train as an up train using the down platform road avoided the signalman having to cross over to the up platform for tablet exchange. Down trains were permitted to use only the down platform. Station staff tended to consist of a station master, two porters and two signalmen, or signalwomen during the Second World War. Some commuters travelled to Bristol while others travelled to Portishead, while some residents of Shirehampton on the opposite bank of the river used the ferry to reach Pill for onwards travel by rail. During part of the 19th century the ferry was owned by Mr and Mrs Porter. His wife checked his takings with a cushion and packet of pins. For each

DESCRIPTION OF THE LINE 87

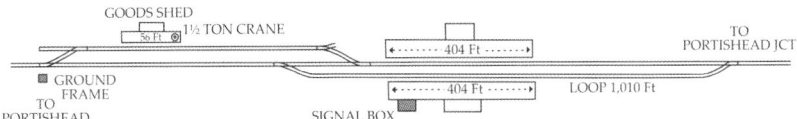

Plan of Pill

Steam railmotor No. 58 arrives at Pill with an up train. *Author's collection*

Pill on 17th February, 1962: '57XX' class 0-6-0PT No. 7729 arrives with the 1.00 pm Saturdays-only Temple Meads to Portishead, crossing dmu Nos. 55032 and 56292 working the 1.15 pm Portishead to Temple Meads. *Hugh Ballantyne*

passenger she spied from the window, she inserted a pin in the cushion and at the end of the day, checked the takings with the number of pins.

Alan Butler, a signalman at Pill, was one of the ferry users. Early one morning he cycled five miles from his home to Shirehampton to catch the ferry over to Pill, but on arrival found that the motor would not start on the regular boat. The ferryman, anxious that Alan should open his box on time, offered to row him across to Pill in a boat half full of muddy water. The River Avon was at flood tide, but with the boatman at the oars, and Alan bailing out water with an old tin, the water level was kept under control and landfall was effected on the far bank in sufficient time for the determined signalman to open his box on time. The ferry closed on 1st October, 1974, having lost traffic due to the opening of the M5 motorway bridge.

Alan Butler recorded some of his other memories at Pill – his job was certainly no sinecure. The distant signals at Pill were over a mile from his box and to move them required great strength and skill due to the weight of wire and pulleys which had to be moved. He braced himself with one foot on the lever frame; released the catch which held the lever secure and with an immense heave and backwards swing pulled the lever towards him. If the lever failed to respond fully to his first swing, the movement needed to be repeated. Replacing a signal involved a pushing movement of the levers.

One Bank Holiday, owing to a signal failure trains could only pass at Pill on the written authority of the signalman. Traffic that day was heavy

Pill goods yard, view down (towards Portishead) *circa* 1960. Note the up home bracket signals in the distance enabling an up train to use either platform.   *M. J. Tozer*

Pill: view west of the station; a new bridge over path/cycle track; the formation of the track to Portishead is seen above. View towards Portbury 29th June, 2001. *Author*

View towards the docks: making an embankment for the new line to Royal Portbury Dock below the M5 bridge which also spans the River Avon, 29th June, 2001. *Author*

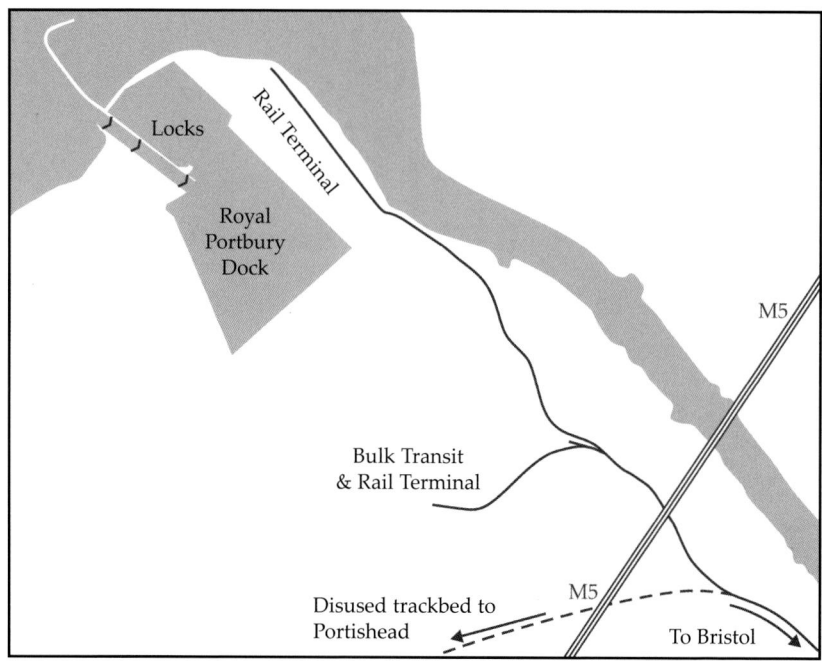

Plan of Royal Portbury Dock.

and Alan Butler ran out of the official forms. Using his initiative he utilized paper from the signal box toilet roll. Passengers leaning out of the carriage windows to see what caused the delay, had the exciting experience of watching the signalman hand a piece of toilet paper to the driver, on receipt of which he blew the whistle, opened the regulator and the train continued on its way!

Beyond Pill the original line falls at a maximum gradient of 1 in 103 to Portbury Shipyard signal box, while a steeply-descending new spur leads from Pill Junction to Royal Portbury Dock (126 miles 58 chains from Paddington). A lake required to be filled in at Portbury to make an embankment and then the lake was required to be replaced with another as it was of special environmental interest. 11,000 trees and shrub were planted to enhance the new section of line.

Beyond Portbury Junction the new line to Royal Portbury Dock passes below the M5 and diverges: a line on the left running to the coal terminal, a cripple siding branching off. The line on the left leads to the car-loading and general cargo sidings. Around the whole site is an embankment to reduce dust being spread and to create a visual barrier.

When the M5 motorway was being constructed in June 1970 a temporary level crossing was made across the Portishead branch at 126 miles 58 chains.

Portbury Shipyard station (127 miles 12 chains) opened 16th September, 1918 and closed 26th March, 1923. The 300 foot long platform and waiting shelter were both of wood. At the west end was a siding 75 feet in length. The station was quite busy with troop trains and those carrying Admiralty civilian personnel. At 127 miles 15 chains a junction offered access four exchange sidings, and following a reversing movement, the shipyard branch. The signal box midway along the sidings opened on 29th January, 1918 and closed 14th April, 1964 as did the 22 chains long passing loop and the two remaining looped sidings, each capable of holding 40 wagons. The crossing loop had only been made available for passenger trains to cross from 2nd April, 1928. The signal box at Portbury Shipyard was large and had 19 levers in a 57-lever frame. As the box had no piped water, this was brought by train. Although situated in an isolated spot, it was a busy box and it was not unknown to have three trains waiting to cross: two passenger and a

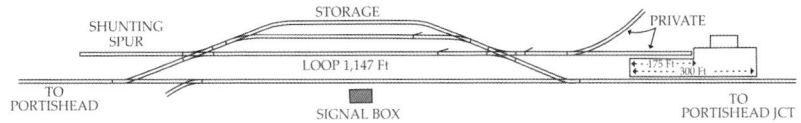

Plan of Portbury Shipyard station

Portbury Shipyard station, view towards Portishead.  *M. J. Tozer*

Ex-GWR diesel railcars W25W and W28W near Portbury Shipyard signal box working the 2.30 pm Temple Meads to Portishead 5th September, 1958.
*Author*

View towards Portishead from Marsh Lane bridge near Portbury Shipyard showing the very overgrown track on 29th June, 2001.
*Author*

goods. Between Portbury Shipyard and Portishead earthworks allow for double track.

In the mid-sixties a liner terminal was proposed at Portbury, the scheme recommended by the National Ports Council, but then a White Paper of 27th July, 1966 said that a case had not been made for it, so the plan was abandoned.

Portbury (127 miles 77 chains) opened 18th April, 1867 and closed to both passengers and goods on 30th April, 1962. When necessary, up to three livestock wagons could be worked from Portishead to Portbury without a brake van, a tail lamp being placed on the last vehicle and a guard or porter accompanying it. The substantial station house still stands. It is believed that the signal box opened in 1880 and closed 10 years later when the up siding was lifted. The single platform was extended from 200 feet to 299 feet in January 1890. At the up end of the station, a short siding worked for most of its life from a ground frame, gave access to two stubs: one serving the carriage dock and the other the cattle dock, respectively holding 3 and 6 wagons. Traffic included cattle,

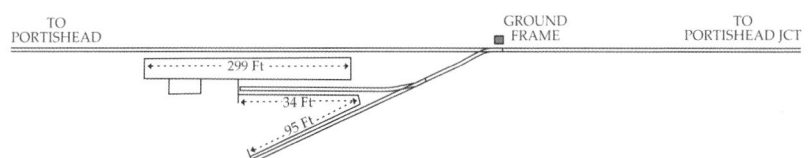

Plan of Portbury station

The single platform at Portbury, view Down.  *Author's collection*

On the flat, straight approach to Portishead the embankment is being widened, 3rd February, 1953, in order to provide a goods loop. *Author's collection*

A culvert is widened, 3rd February, 1953. *Author's collection*

horse boxes and pit props. In the 1950s there was an attractive garden opposite the platform. Tickets of passengers to Portishead were collected at Portbury.

An interesting feature opposite Portbury station was a Bristol & Exeter Railway carriage constructed in 1863 and condemned in 1890 where it became a Wesleyan Mission Room. Following the completion of a permanent building, E. H. Shopland from Clevedon and his horse Smart dragged the coach along the road on rollers to stand beside the new building where it remained until December 1969.

Originally the level line curved beyond Portbury and between the gas works and the passenger station, crossed Portishead Pill by the curving, timber Portishead viaduct with 23 spans, the underside being about 5 feet above the water. This timber viaduct at Blackwater was to have been replaced by a steel bridge in 1919 and although the steelwork was actually delivered, it was replaced by a culvert consisting of a 3-foot diameter concrete pipe in 1920.

Originally all the railway lines were on the west side of Portishead dock, but over the years a complex developed on both sides. Portbury Timber Jetty ground frame (129 miles 32 chains) was brought into use 22nd April, 1903 and moved a short distance 13th February, 1928. It served Severn Kraft Mills Limited which went into liquidation in June 1934, but the sidings were allowed to remain until June 1944. The same ground frame controlled access to Bristol Corporation's sidings under an

The Permanent Way Construction brake van from St Philip's Marsh at Portishead, 3rd February, 1953. *Author's collection*

The approach to Portishead, view east from the water tank, 9th March, 1954.
*Author's collection*

agreement of 11th March, 1903. These served an extensive timber wharf and associated yards and sheds on the east side of the dock. When a BP oil depot was opened about 1908, access was also from this ground frame. Still nearer Portishead was the Gas Siding ground frame opened in 1877 and the siding lifted in August 1950. Portishead viaduct was shortened and then eventually the land raised eliminating it completely.

The line curved round the southern end of the dock to the original single 408 foot-long platform passenger station at Portishead (129 miles 73 chains) opened 18th April, 1867 and it, together with the second platform opened 9th March, 1930, closed 4th January, 1954. The substantial brick building contained the station offices and the station master's residence, a canopy shielded the platform. The licensed refreshment room in addition to serving passengers, was used by other local residents, dockers and power station workers. On 9th March, 1930 an up platform economically constructed from timber and cinders, was brought into use alongside the run-round loop. The new platform did not receive a canopy until 1948. In 1930 the station had 25 employees. In 1956 Portishead had a 6-ton crane. The station was particularly busy on bank holidays and a curiosity was that more people would return than arrived. This was because quite a few passengers who had arrived at Portishead by bus found them full for the the return journey and so used the train.

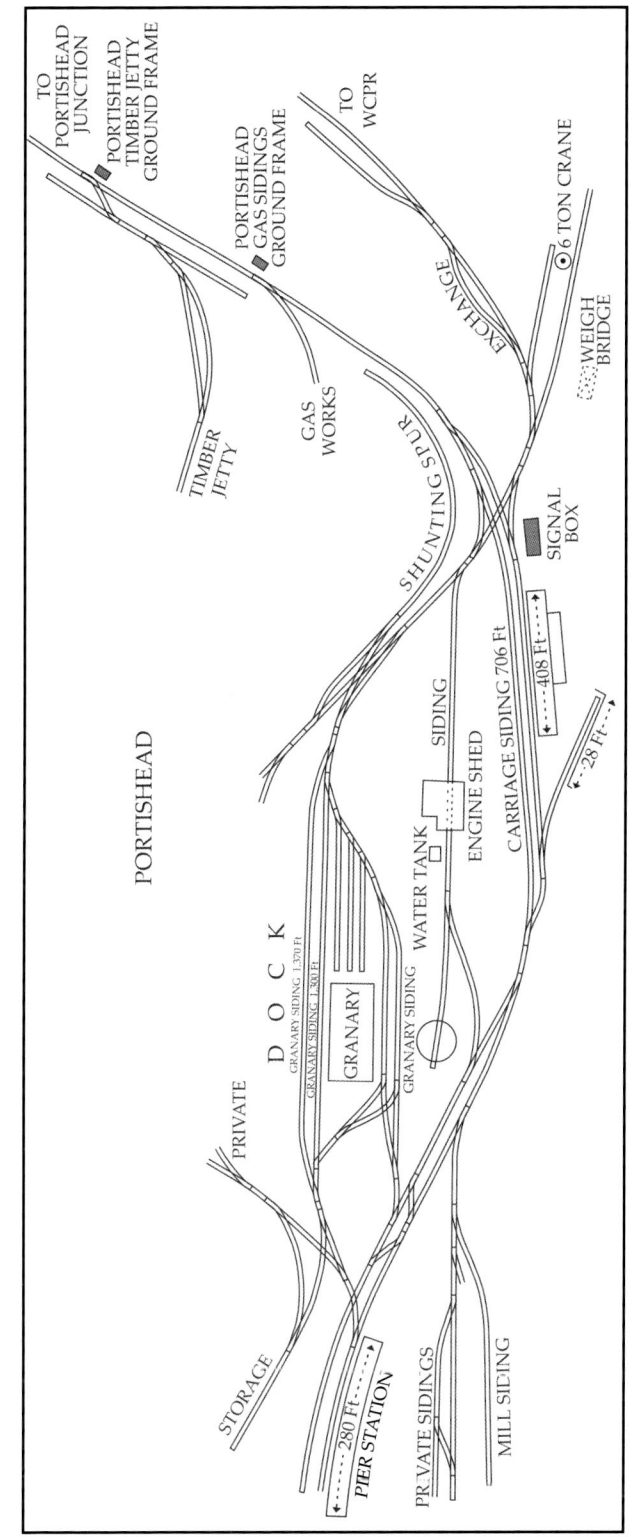

Plan of Portishead (pre-1954).

The carriage approach to Portishead station *circa* 1905.  *Author's collection*

The exterior of Portishead station *circa* 1905.  *Author's collection*

Portishead station view towards the pier *circa* 1910; the engine shed is on the right.
*Author's collection*

The power station chimneys dominate Portishead station and signal box, 9th April, 1952.
*Dr A. J. G. Dickens*

Portishead station, view down 9th April, 1953. *Dr A. J. G. Dickens*

Portishead station, view south (towards Bristol) 4th January, 1954. The last day of the old station. *Dr A. J. G. Dickens*

An ex-GWR diesel railcar forms the last train from the old Portishead station, 4th January, 1954. Note the wartime camouflage painted on the chimney of the power station.
*Michael Farr*

Portishead turntable and sidings 15th May, 1952.      *Author's collection*

The original stone and timber signal box at Portishead. The original photograph was postally franked on 24th August, 1904. It was replaced on the same site by a larger brick-built box 16th June, 1908. *M. J. Tozer*

The brick-built replacement box at Portishead, opened 16th June, 1908. *M. J. Tozer*

Portishead signal box, immediately south of the passenger station, under an agreement of 11th August, 1908 gave access to the exchange sidings with the Weston, Clevedon & Portishead Railway. These sidings were brought into use on 7th August, 1907 and first used on 2nd November, 1908. When the light railway closed on 18th May, 1940 the sidings were retained. The key to the exchange sidings was kept in Portishead signal box and neither locomotives of the Bristol Docks Committee or the Weston, Clevedon & Portishead Light Railway were allowed into the sidings.

Portishead 'A' Power Station to the west of the passenger station, opened on 25th March, 1929, the associated sidings being brought into use 1st March, 1929. In addition to bringing freight and mineral traffic to the branch, it also brought in workers.

The pier at Portishead opened in June 1868 and the low water extension of 300 feet on 18th April, 1870, the latter enabling the ferry service to be operated at any state of the tide. Daily steamer services to Newport and Cardiff were operated by the railway while in summer there were trips to Ilfracombe with through rail and steamer bookings from both the GWR and the Midland Railway (MR) and until the opening of the line to Ilfracombe on 20th July, 1874 the GWR and the MR routed passengers for Ilfracombe via Portishead and a steamer. The opening of the Severn Tunnel to passengers on 1st December, 1886 marked the end of the cross-channel ferry service.

Portishead Pier station (130 miles 29 chains), opened in the late 1870s, had a single 280-foot long platform and run-round loop. It is believed that it was used almost entirely for freight traffic, any passenger trains ceasing to use it after 1884 when regular steamer services stopped calling at Portishead Pier. Certainly in the 1930s passenger traffic was prohibited from using the line. The station building was not demolished until 1954 during the construction of the Portishead 'B' Power Station. A line branched north past extensive cattle pens to run along the actual pier.

On the east side of the dock, about half of the timber yards was taken over by Albright & Wilson's Works the first traffic arriving on 22nd February, 1954 and the last traffic leaving in 1978.

The Mustad nail factory made use of the railway, wire arriving from Roath Docks, Cardiff and 28-pound boxes of nails being dispatched. With closure of the Weston, Clevedon & Portishead Light Railway in 1940, stone which had hitherto travelled by rail from Black Rock Quarry to the GWR, was brought by road to the GWR railhead at Portishead.

104  THE BRISTOL TO PORTISHEAD BRANCH

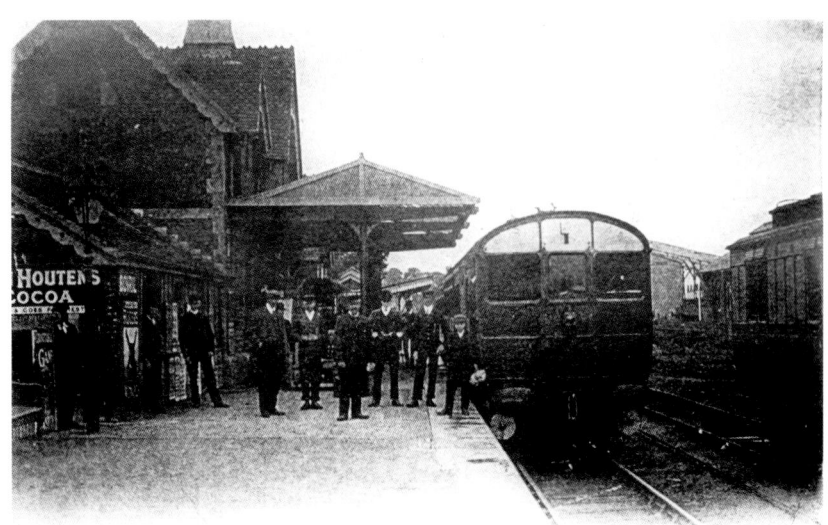

A steam railmotor at Portishead. *Author's collection*

Portishead *circa* 1910 showing the goods yard and a ship in the dock. The turntable is just beyond the shed in the centre foreground. The passenger station is on the right.
*Author's collection*

Portishead, new station site, 30 foot deep piles. 23rd April, 1953. The chimneys of
Portishead 'A' Power Station dominat the background.  *BR*

The construction of Portishead 'B' Power Station on a site between 'A' Power Station and the docks required the demolition of the original station and the construction of a new terminus. Following the departure of the 9.05 am Portishead-Bristol hauled by 2-6-2T No. 5528 on 4th January, 1954 to the sound of detonators, staff transferred from the old to the new station about half a mile distant where the first train to arrive was the 12.11 pm from Temple Meads.

The new station, said to be the first new station to be built in England after the Second World War, was a modern design by H. E. B. Cavenagh in concrete and limestone and built at a cost of £250,000 paid for by the Electricity Generating Board. The two-faced platform built on piles, was 35 feet wide and 754 feet in length, largely covered by a 600 foot-long flat roof of concrete and glass. An unconventional construction method was used for the roof canopy formed of post-tensioned beams and prestressed decking slabs, prefabricated to facilitate speedy and economical erection. The platform was illuminated by 70 bowl lamps. The waiting rooms had the contemporary colour scheme of contrasting white and lemon-yellow distempered walls, rubber tiled floors and light

The exterior of Portishead station, 5th September, 1964.  *Michael Farr*

coloured furniture suggestive of Swedish design. Two run-round lines were provided and two carriage sidings capable of holding 10 70-foot coaches..

The station entrance set back about 50 feet from the main road on a semi-circular drive, had swing glass doors leading into the booking hall. Situated within yards of the main shopping street its location was ideal for the scores of business people and holidaymakers who found it easier to travel by rail.

Passengers welcomed the new station as it had better facilities – *The Architect* for March 1954 commented on the comfortable seats – and passenger traffic increased. Adjacent to the new station was a 10-road goods yard with a capacity for 300 wagons.

The new passenger station (129 miles 64 chains) had been built on reclaimed marsh land well covered with thousands of tons of power station ash, so despite the fact that it was on 30 supporting piles 56-40 feet deep, soon settlement problems were experienced. When the author saw the station on 3th June, 1963, platform 2 was closed to the public and the asphalt in the centre of the platform had risen.

The *Bristol Evening Post* for 27th August, 1964 painted a gloomy picture:

> The station which cost thousands has been allowed to decay. Entrance gates and doors which once shone now bear the scuff marks of time. Its ticket office walls show streaky soot marks from the familiar stoves, and everywhere shabbiness has taken hold.

Moving from the old Portishead station to the new, 4th January, 1954. The staff, left to right are: G. Richards, station master; Peter Stonestreet, lorry driver; Alex Shoreland; Albert Spencer, porter; Cyril Hayman (carrying box); Tom Pugsley (behind Cyril Hayman); Fred Willmott and Norman Jelley. Several years later, Peter Stonestreet had promotion to mobile crane driver in charge of a Ransome's solid-tyred Iron Fairy, 5-ton capacity with rear steering. As required he travelled to various stations in the district.  *Author's collection*

Entrance to the new Portishead station, 30th December, 1953.  *BR*

Platform No. 1 *circa* 1953, ground frame and run-round loop on the left.   *Lens of Sutton*

Portishead, view to buffers, 20th August, 1960.   *R. E. Toop*

DESCRIPTION OF THE LINE 109

No. 1 platform, view to buffers *circa* 1962. *Lens of Sutton*

View up 7th February, 1962, towards the water tower. *BR*

Six-wheel Signal & Telegraph mess coach, 17th February, 1954.   *BR*

View Up from water tower, 1st October, 1953.   *BR*

Due to subsidence No. 2 platform is closed to the public. *Lens of Sutton*

Only the waiting rooms gleam from behind locked doors; their tables and chairs neatly arranged – unavailable to travellers who must ask for them to be unlocked if they wish to use them. Here and there light bulbs point from shadeless fittings and paint blisters beneath concrete beams, where friendly sparrows hop about the eaves.

In the entrance hall a crudely lettered poster announces the departure times of the six daily trains to Bristol; next to it a pitted blackboard bears the chalk-written death sentence. [This was probably a management ploy not to supply a printed departure sheet in order to deter potential passengers and thus justify claims for closure. The station staff did its best to encourage traffic and so provided a handwritten notice.]

In the seven months of construction it was not without its problems. It was impossible to get a mason to tackle the undressed rubble and finally a wall mason undertook to train bricklayers. By the completion, eight untrained men had qualified as masons.

Passenger trains are limited to six in each direction and the illogical timing of these has driven much potential traffic on to the roads.

At 3s. return, the fare is still a shilling less than the bus fare and with the running time of 28 minutes compared to the bus's 55 minutes, the rail service should be the logical answer to those who seek a speedy service. But with falling receipts -the line carries about 450 in each direction a week – the death knell was rung for the line.

The new signal box in yellow brick, had a roof of aluminium sheeting from which was suspended an acoustic ceiling to absorb noise found in a manually-operated signal box and to give added thermal insulation. and continuous windows on three sides canted at an angle of 10 degrees, thus throwing all reflected glare up to the ceiling. The lever frame on the blind wall had 47 working levers, 24 spares and 12 spaces. The lines were track-circuited as far as the down distant signal set 1,923 yards from the box. The signalman had the luxury of an indoor water closet and did not need to brave the elements. Interior lighting was by low-intensity spot lights directed on the instruments and track diagram, while diffused lighting was used for the remainder of the box.

The *Bristol Evening Post* dutifully recorded details of the last train:

A kiss awaited for Driver Ray Mitchell when he piloted the 6.45 Bristol-Portishead train into the north Somerset coastal town on Saturday. It was planted on his cheek by Mrs Edith Hunt of 17, High Street, Portishead.

Driver Mitchell – with footplate colleagues Driver Doug Evans and Fireman Peter Sheppard – had just driven the last passenger train from Bristol to Portishead. Mrs Hunt was one of the crowd – the largest seen on the station for years – that turned out to welcome it.

There was a carnival atmosphere about this last round trip between Temple Meads and Portishead. Upwards of 150 rail enthusiasts boarded the train at the start and at each halt along the 11-mile journey more piled aboard with their note-books, cameras and tape recorders.

Too young to be a rail enthusiast was the youngest passenger...15-month-old Michael Foss. His parents, Mr and Mrs Alan Vincent Foss, of Slade Road, Portishead, wheeled him aboard – still in his pram – at Pill. "This is no special journey for us," said Mr Foss, "We always catch this train after Saturday visits to Pill, but we won't be able to do it any more."

A quick shunting operation at Portishead, a scramble for seats, a wave from relief station master Wilfred Cooling and the return journey was under way. For many it was a standing room only trip, but no one seemed to care.

With passengers cheering at every opportunity, detonators at every station signalling the end of the service and line-side householders braving the evening rain to wave farewell, the train steamed out of the railway time-tables.

At Temple Meads the fans crowded round to make the last notes in their logs. Flash-bulbs popped, tape recorders whirred and the train crew took the carriages off to the sheds. It was exactly 8.14 pm by the station clock when the last lights went out on the Bristol-Portishead passenger line.

Following the station's closure it became a petrol filling station.

New signal box opened 4th January, 1954, seen here on 1st October, 1953. The signal on the right was one of the few of the upper-quadrant type installed on the Western Region. *BR*

Interior of new signal box, 1st October, 1953. Note the central heating radiator on the right. *BR*

Diesel multiple-unit B575, class '119' built by the Gloucester Railway Carriage & Wagon Company, Nos. W51060/59419/51088, chartered by Gordano School to carry pupils from Portishead to Bath Spa on 7th October, 1977, 14 years after the cessation of regular passenger trains on the branch. Here it is seen waiting in the siding at Portishead.   *Michael Farr*

The dmu chartered by Gordano School from Portishead to Bath Spa, near the former Portbury station.   *Michael Farr*

Outward and return tickets issued for the Portishead to Bath Spa trip.  *Michael Farr*

Cover carried by Gordano's School's special train.  *Michael Farr*

Carried on
Special
Passenger Train

TRAVEL CENTRE
-7 OCT 1977
BATH SPA

PORTISHEAD

TO

BATH

7th October 1977

Mr. & Mrs. Michael Farr,
54, Godwin Drive,
Nailsea,
BRISTOL.            BS19  2XE

'Dean Goods' 0-6-0 No. 2426 (82A Bristol, Bath Road shed) at Ashton Carriage Sidings in the early fifties. *Author's collection*

A dmu working the 4 30 pm Temple Meads to Portishead calls at Portbury on 26th April, 1962, two days before the last passenger train at that station. *E. T. Gill*

# Chapter Ten

## Locomotives & Coaches

It is not known what locomotives first worked the branch, but in its early years certainly a 0-6-0ST was used and by the 1880s the '517' class 0-4-2Ts were working the branch, No. 1472 being recorded on 23rd August, 1895 and No. 1479 the following day. On Easter Monday 1900, the branch was worked by Armstrong's double-framed 0-6-0STs of the '1600' or 'Buffalo' class, built between 1874 and 1881.

Almost opposite the passenger platform at Portishead was the 50 foot by 24 foot single-road engine shed with brick walls and a gable style slated roof. Adjacent was the office, stores and coal plant, with the water tank above the pump house, a little further north. In June 1896 the 45-foot diameter turntable adjacent was moved further north to about 200 yards from the north end of the station. The actual rail length on the table was 44 feet 7 inches and extension rails were fitted to enable those locomotives with a wheelbase of 50 feet 3 inches to be turned, this facility being used for Standard Goods class 0-6-0s. The turning mechanism was operated from the table itself. Also about 1896 the coaling road and shelter were dismantled, certainly they were out of use by 1901. Since at least 1896 no engine had been allocated to the shed and a note dated 18th April, 1903 stated that the 'shed remains open for the purposes of access to turntable and running round trains'. Some of the steam railmotors shedded at Bristol between May 1909 and October 1935 worked the Portishead branch. 'Dean Goods' 0-6-0s were seen on some passenger trains. During the Second World War at least one LMS 0-6-0 on loan to the GWR worked over the branch.

In the 1950s local passenger traffic was generally worked by '45XX' and '4575' class 2-6-2Ts, the '4575' variety preferred as they had a water capacity of 1,300 gallons rather than just 1,000. Other locomotives used on passenger and goods duties included '27XX',

| From PORTISHEAD | | |
|---|---|---|
| 6 44 | Bristol (T.M.).. | .. D.C.28 |
| 7 53 | Bristol (T.M.).. | .. D.C.29 |
| 8 18 | Bristol (T.M.).. | .. 458 |
| 9 0 | Bristol (T.M.).. | .. D.C.28 SX, 485 SO |
| 10 15 | Ashton Gate SO | .. D.C.29 |
| 10 45 | Ashton Gate SO | .. 488 |
| 11 15 | Bristol (T.M.) SX | .. D.C.28 |
| 11 45 | Bristol (T.M.) SO | .. 488 |
| 12 15 | Bristol (T.M.) SO | .. D.C.29 |
| 1 15 | Bristol (T.M.).. | .. 460 SO, D.C.28 SX |
| 1 45 | Bristol (T.M.) SO | .. 517 |
| 2 15 | Bristol (T.M.) SX | .. D.C.29 |
| 2 15 | Ashton Gate SO | .. D.C.29 |
| 3 15 | Bristol (T.M.).. | .. 460 SO, D.C.28 SX |
| 4 15 | Bristol (T.M.).. | .. D.C.29 |
| 5 15 | Bristol (T.M.).. | .. 517 |
| 5 47 | Bristol (T.M.) SX | .. D.C.28 |
| 6 15 | Bristol (T.M.).. | .. 518 |
| 7 15 | Bristol (T.M.).. | .. 466 SO, D.C.28 SX |
| 8 15 | Bristol (T.M.).. | .. 468 SO 519 SX |
| 10 15 | Bristol (T.M.) SO | .. 458, 484 |

Weekday working of diesel railcars 15th June, 1959 – 13th September, 1959. DC stands for diesel car, the 3 figure number is the set number of coaching stock.

BR Standard class '3' 2-6-2T No. 82040 at Portishead with the 12.15 Saturdays-only to Temple Meads, 20th August, 1960.
*R. E. Toop*

Ex-GWR diesel railcar W28W at Portishead with the 1.45 pm to Temple Meads, 27th August, 1954.
*Hugh Ballantyne*

'20XX', '36XX', '77XX', '84XX', '87XX', '94XX' class 0-6-0PTs, and the Standard Goods '2301' class, BR Class '3' 2-6-2Ts appeared and single or 3-car sets of ex-GWR diesel railcars, 0-6-0PTs appearing on freights.

'Dotted Red' classification being applicable Parson Street Junction-Clifton Bridge, all classes of locomotive were permitted except a 'King', '47XX' 2-8-0, a class '8' 4-6-2 and a gas turbine. A 'West Country' class 4-6-2 was allowed to work to Clifton Bridge without restriction, but a 'King Arthur' class 4-6-0 was limited to 20 mph. Beyond Clifton Bridge only 'Yellow' classes were allowed, although 'Red' 94XX class 0-6-0PTs were permitted. In 1964 the '82XXX' class '3MT' 2-6-2Ts were authorized to work to Portishead.

'57XX' class 0-6-0PTs could work Temple Meads-Ashton Swing Bridge North via Wapping and also Canon's Marsh – Ashton Junction.

All up double-loaded freight trains were required to be assisted up the 1 in 200 gradient between Ashton Junction and Parson Street Junction by an engine in the rear.

In the early 1950s when Bristol City were playing at home, a 'Castle' class 4-6-0 sometimes worked a 10-coach train which originated at Henbury and called at intermediate stations via Filton to terminate at Ashton Gate. Through football specials from the Southern Region were generally hauled by Pacifics.

*Circa* 1964 when steam engines were being withdrawn, some classes were classified UBA – 'Use to Best Advantage' and 4-6-0 'Castles', 'Halls', ex-GWR and LMS 2-8-0s and 2-10-0 class '9's were used on Albright & Wilson's phosphate trains from Portishead.

On 3rd January, 1955 regular interval services were instituted on Bristol suburban lines for which three single-unit ex-GWR railcars were allocated to Bristol. At the same time, the twin unit was transferred to the Avonmouth and Portishead services.

On 17th November, 1958 the Avonmouth service was completely dieselized with BR dmus and on the same date the ex-GWR railcars were confined to Portishead and Bath Green Park workings. Portishead services were worked almost entirely by ex-GWR railcars; one set consisting of a 3-car unit comprising two single cars with a BCK (brake corridor composite) coach sandwiched between them and the other a single car. Despite some early doubts about clearances of BR Derby suburban 3-car dmu 63-foot 6 inch stock on the Portishead branch when substituted for a failed ex-GWR railcar, no trouble was experienced and the sets were then used extensively on the line. The 11th April, 1959 saw the debut on the branch of a Swindon Works Cross-Country dmu.

In the summer of 1961 some Portishead trains were actually advertised as diesel workings, but hitherto, although some had been worked by BR dmus, the local time table had shown all trains as steam-hauled, due to the

A 2-6-2T being watered at Portishead.  *Author's collection*

Hymek D7023 at Portishead with the 17.15 to Temple Meads, 28th August, 1964.

*E. T. Gill*

the Western Region's policy of not advertising diesel services on branches which were 'under review'. Two Pressed Steel diesel single cars with drive-end trailers were allocated to work some of the branch trains.

Goods engines generally worked bunker-first to Portishead and chimney-first to Bristol. But there was no standard practice for passenger locomotives, engines working either way in both directions.

The engine which worked the last down passenger train left its coaches at Portishead before returning light to Bath Road shed. Having worked all day by the time she was screwed down on the coal road her fire was blue with clinker and there was little more than coal dust remaining in her bunker. The locomotive heading the first down goods the following morning returned with the first up passenger train. The coaching stock was supplied from Dr Day's sidings (Bristol).

SPEED RESTRICTIONS
An overall speed restriction of 40 mph
Clifton Bridge 10 mph
Oakwood at end of up loop 20 mph.
Pill 10 mph
Portbury Shipyard up loop 10 mph

With the opening to Royal Portbury Dock, an overall speed restriction of 30 mph applied.

**Wapping Wharf** (*see chapter 15*)
No. 1 engine worked 144 hours per week 6.00 am Monday – 6.00 am Sunday continuously.
No. 2 engine worked 15½ hours per week Mondays – Fridays 4.45 pm – 7.50 pm (No. 6 Transfer).
No. 3 engine worked 5½ hours per week Saturdays-only 1.30 pm – 7.00 pm.

**Canon's Marsh** (*see chapter 17*)
No. 1 engine worked 84 hours 6 .00 am – 8.00 pm daily.
No. 2 engine worked 35 hours 5.00 pm – 10.00 pm daily.

**Coaches**
Certainly by April 1918 one Clifton Downs auto set of low-roofed coaches made a daily return trip to Portishead.

In the 1950s passenger trains were generally of four to six coaches of mixed suburban and corridor stock.

# BRISTOL & PORTISHEAD PIER & RAILWAY.

# TIME TABLE FOR SEPTEMBER, 1869.

## DOWN TRAINS.

| FARES Ordinary 1st s. d. | Ordinary 2nd s. d. | Ordinary 3rd s. d. | Return 1st s. d. | Return 2nd s. d. | STATIONS. | Week Days 1 2 3 Class. | 1 2 3 Class. | 1 2 3 Class. | 1 2 3 Class. | 1 2 3 Class. | 1 2 3 Class. | 1 2 3 Class. | 1 2 3 Class. | Sundays 1 2 3 Class. | |
|---|---|---|---|---|---|---|---|---|---|---|---|---|---|---|---|
| | | | | | | A.M. | A.M. | A.M. | P.M. | P.M. | P.M. | P.M. | P.M. | A.M. | |
| | | | | | London .. .. .. .. dep. | .. | .. | 6 0 | 9 15 | 11 45 | .. | 2 0 | 4 50 | 10 0 | .. |
| | | | | | Bath .. .. .. .. .. .. | .. | 7 0 | 9 45 | 12 0 | 2 15 | 3 50 | 5 20 | 7 40 | P.M. 2 49 | .. |
| | | | | | BRISTOL (B. & E. Station)... | 6 5 | 8 10 | 10 20 | 12 45 | 3 10 | 5 10 | 6 50 | 8 35 | 3 40 | ... |
| 0 6 | 0 4 | 0 3 | 0 9 | 0 6 | Clifton Bridge ... ... ... | 6 20 | 8 21 | 10 30 | 12 57 | 3 21 | 5 21 | 7 0 | 8 46 | 3 52 | ... |
| 1 3 | 0 10 | 0 7½ | 2 0 | 1 3 | Pill ... ... ... ... ... ... | 6 40 | 8 33 | 10 41 | 1 11 | 3 33 | 5 33 | 7 12 | 8 58 | 4 6 | ... |
| 1 7 | 1 1 | 0 9½ | 2 4 | 1 6 | Portbury ... ... ... ... | 6 45 | 8 39 | ... | 1 16 | 3 39 | 5 39 | 7 18 | 9 4 | 4 11 | ... |
| 2 0 | 1 6 | 0 11½ | 3 0 | 2 3 | PORTISHEAD... ... ... arr. | 6 50 | 8 45 | 10 50 | 1 25 | 3 45 | 5 45 | 7 25 | 9 10 | 4 20 | ... |

## UP TRAINS.

| FARES Ordinary 1st s. d. | Ordinary 2nd s. d. | Ordinary 3rd s. d. | Return 1st s. d. | Return 2nd s. d. | STATIONS. | Week Days 1 2 3 Class. | 1 2 3 Class. | 1 2 3 Class. | 1 2 3 Class. | 1 2 3 Class. | 1 2 3 Class. | 1 2 3 Class. | 1 2 3 Class. | Sundays 1 2 3 Class. | |
|---|---|---|---|---|---|---|---|---|---|---|---|---|---|---|---|
| | | | | | | A.M. | A.M. | A.M. | P.M. | P.M. | P.M. | P.M. | P.M. | P.M. | |
| | | | | | PORTISHEAD ... ... dep. | 7 15 | 9 0 | 11 5 | 1 45 | 4 5 | 6 0 | 7 45 | 9 25 | 8 30 | ... |
| 0 5 | 0 3 | 0 1¼ | 0 8 | 0 5 | Portbury... ... ... ... | 7 20 | 9 5 | ... | 1 50 | 4 10 | ... | 7 50 | 9 30 | 8 36 | ... |
| 0 9 | 0 6 | 0 3½ | 1 2 | 0 9 | Pill ... ... ... ... ... | 7 26 | 9 11 | 11 14 | 1 55 | 4 16 | 6 8 | 7 56 | 9 36 | 8 42 | ... |
| 1 6 | 1 0 | 0 8 | 2 3 | 1 6 | Clifton Bridge ... ... ... | 7 40 | 9 25 | 11 28 | 2 10 | 4 30 | 6 20 | 8 10 | 9 50 | 8 58 | ... |
| 2 0 | 1 6 | 0 11½ | 3 0 | 2 3 | BRISTOL (B. & E. Station) arr. | 7 50 | 9 35 | 11 40 | 2 20 | 4 50 | 6 30 | 8 20 | 10 0 | 9 10 | ... |
| | | | | | Bath .. .. .. .. .. .. | 8 17 | 10 25 | 12 2 | 3 2 | 5 45 | 7 17 | 9 15 | 1 8 | 9 15 | .. |
| | | | | | London .. .. .. .. .. | 11 15 | 2 30 | 2 45 | 6 0 | .. | 10 15 | .. | 4 35 | 4 35 | .. |

**Steamers call (weather permitting) at or off PORTISHEAD PIER on their voyages between Bristol and Bideford, Barnstaple, Cardiff, Carmarthen, Cork, Dublin, Hayle, Ilfracombe, Liverpool, Newport, Neath, Padstow, Swansea, Tenby, Waterford and Wexford.**

**Steamers ply Daily between the PIER at PORTISHEAD and ILFRACOMBE (Fares 3s and 5s), and CARDIFF and NEWPORT (Fares 10d and 1s 6d).**

**PERIODICAL TICKETS will be issued on the following terms:—**

| Between | One Month 1st Cls. £ s. | One Month 2nd Cls. £ s. | Two Months 1st Cls. £ s. | Two Months 2nd Cls. £ s. | Three Months 1st Cls. £ s. | Three Months 2nd Cls. £ s. | Six Months 1st Cls. £ s. | Six Months 2nd Cls. £ s. | Twelve Months 1st Cls. £ s. | Twelve Months 2nd Cls. £ s. |
|---|---|---|---|---|---|---|---|---|---|---|
| Bristol & Portishead ... ... | 2 0 | 1 12 | 3 10 | 3 0 | 5 0 | 4 4 | 8 0 | 7 0 | 12 12 | 11 0 |
| Clifton Bridge " ... ... | 1 10 | 1 5 | 2 14 | 2 5 | 3 10 | 3 0 | 6 6 | 5 5 | 10 10 | 9 0 |

The Telegraphic Communications on this Line are now open to the public for transmission of Home or Foreign Messages.

### PARCEL RATES.

| | s. d. | | s. d. |
|---|---|---|---|
| Not exceeding 7 lbs. ............ | 0 4 | Not exceeding 56 lbs. ............ | 0 10 |
| 14 lbs. ............ | 0 6 | " 84 lbs. ............ | 1 0 |
| 28 lbs. ............ | 0 8 | " 112 lbs. ............ | 1 2 |

One Penny for every additional 8lbs. All fractional parts will be charged as 8 lbs.

ARROWSMITH, Printer, 11, Quay Street, BRISTOL

Timetable for September 1869.

Branch Working Timetable October 1886



## PORTISHEAD BRANCH. Narrow Gauge.

The Line from Bristol to Clifton Bridge is double, and from Clifton Bridge to Portishead single.

**The Train Staff Stations are Clifton Bridge, Pill and Portishead. Pill is the intermediate crossing place.**

| Section. | Shape of Staff and Ticket. | Colour of Staff and Ticket. |
|---|---|---|
| Clifton Bridge and Pill .. .. .. .. | Triangular. | White. |
| Pill and Portishead .. .. .. .. | Round. | Blue. |

**Down Trains. BRISTOL TO PORTISHEAD.**

| Miles from Bristol. | STATIONS. | 1 Goods | 2 Pass. | 3 Cond'l Goods | 4 Pass. | 5 Cond'l Goods | 6 Pass. | 7 Goods | 8 Pass. | 9 Cond'l Goods | 10 Pass. | 11 Pass. | 12 Cond'l Goods | 13 Pass. | 14 Cond'l Goods | 15 Pass. | 16 Pass. | 17 | 18 | Sundays 1 Pass. | 2 |
|---|---|---|---|---|---|---|---|---|---|---|---|---|---|---|---|---|---|---|---|---|---|
| | | A.M. | A.M. | A.M. | A.M. | A.M. | A.M. | A.M. | P.M. | P.M. | P.M. | P.M. | P.M. | P.M. | P.M. | P.M. | P.M. | | | P.M. | |
| — | Bristol .. dep. | .. | 6 55 | .. | 7 45 | .. | 10 10 | .. | 12 45 | .. | 2 45 | 3 50 | .. | 5 25 | .. | 7 20 | 8 30 | .. | .. | 3 45 | .. |
| ¾ | Pyllo Hill " | 5 45 | — | 7 0 | — | 9 20 | — | 10 50 | — | 1 30 | — | — | 4 50 | — | 6 10 | — | — | .... | .... | — | .... |
| 1 | Bedminster " | — | — | — | 7 50 | — | 10 15 | — | — | — | — | 3 55 | — | — | — | 7 25 | 8 35 | .... | .... | — | .... |
| 1½ | Malago Siding " | — | — | RR | — | RR | — | — | — | RR | — | — | RR | — | RR | — | — | .. | .. | — | .. |
| 2 | Portishead Jun.," | 5 50 | 7 0 | 7 5 | 7 55 | 9 25 | 10 20 | 10 55 | 12 50 | 1 35 | 2 50 | 4 0 | 4 55 | 5 30 | 6 15 | 7 30 | 8 40 | .... | .... | 3 50 | .... |
| 3 | Ashton Siding " | CR | — | — | — | — | — | CR | — | — | — | — | — | — | — | — | — | .... | .... | — | .... |
| 3¾ | Clifton Bridge " | 6 5 | 7 5 | 7X30 | 8 0 | CS | 10 25 | 11X29 | 12X55 | 2X 0 | 2 55 | 4 5 | 5X 5 | 5 35 | CS | 7 35 | 8 45 | .... | .... | 3 57 | .... |
| 7¾ | Pill .... " | 6 20 | 7X17 | CS | 8X12 | CS | 10X37 | 11 50 | 1 7 | 2X15 | 3 7 | 4X17 | 5 20 | 5 47 | 7X 0 | 7X47 | 8X57 | .. | .. | 4 11 | .. |
| 9¾ | Portbury .. " | 6 35 | 7 25 | — | 8 20 | — | 10 45 | CR | 1 15 | — | 3 15 | 4 25 | — | 5 55 | — | 7 55 | 9 5 | .... | .... | 4 20 | .... |
| 11¾ | Portishead arr. | 6 45 | 7 30 | 7 55 | 8 25 | 10 5 | 10 50 | 12 5 | 1 20 | 2 30 | 3 20 | 4 30 | 5 30 | 6 0 | 7 15 | 8 0 | 9 10 | .... | .... | 4 25 | .... |

**Up Trains. PORTISHEAD TO BRISTOL.**

| Miles from Portishead | STATIONS. | 1 Pass. | 2 Pass. | 3 Pass. | 4 Cond'l Goods | 5 Pass. | 6 Cond'l Goods | 7 Pass. | 8 Goods | 9 Pass. | 10 Cond'l Goods | 11 Pass. | 12 Pass. | 13 Cond'l Goods | 14 Pass. | 15 Goods | 16 Cond'l Goods | 17 | 18 | Sundays 1 Pass. | 2 |
|---|---|---|---|---|---|---|---|---|---|---|---|---|---|---|---|---|---|---|---|---|---|
| | | A.M. | A.M. | A.M. | A.M. | A.M. | P.M. | P.M. | P.M. | P.M. | P.M. | P.M. | P.M. | P.M. | P.M. | P.M. | P.M. | | | P.M. | |
| | Portishead dep. | 7 5 | 8 0 | 8 55 | 10 20 | 11 5 | 12 15 | 1 40 | 1 45 | 3 30 | 4 0 | 4 40 | 6 45 | 7 30 | 8 45 | 9 40 | 9 55 | .... | .... | 8 15 | .... |
| 2 | Portbury " | 7 10 | 8 5 | 9 0 | — | 11 10 | — | — | 2 0 | 3 35 | — | 4 45 | 6 50 | — | 8 50 | — | — | .. | .. | 8 20 | .. |
| 4 | Pill .. " | 7X17 | 8X16 | 9 7 | 10X38 | 11 17 | CS | 1 48 | 2X15 | 3 42 | 4X18 | 4 52 | 6X57 | 7X48 | 8X58 | 10 0 | CS | .... | .... | 8 25 | .... |
| 8¼ | Clifton Bridge " | 7X29 | 8 28 | 9 17 | CS | 11X27 | 12X50 | 1X59 | 2 30 | 3 54 | CS | 5X 4 | 7 8 | 8 10 | 9 10 | 10 20 | CS | .. | .. | 8 38 | .. |
| 8¾ | Ashton Siding " | — | — | — | — | — | — | — | CR | — | — | — | — | — | — | — | — | .. | .. | — | .. |
| 9¾ | Portishead Jun.," | 7 32 | 8 31 | 9 21 | 11 10 | 11 32 | 1 5 | 2 4 | 2 45 | 3 58 | 4 50 | 5 10 | 7 11 | 8 20 | 9 14 | 10 35 | 10 50 | .... | .... | 8 43 | .... |
| 10¼ | Malago Siding " | — | — | — | RR | — | RR | — | CR | — | RR | — | — | RR | — | — | RR | .. | .. | — | .. |
| 10¾ | Bedminster " | 7 37 | 8 36 | 9 25 | — | 11 37 | — | 2 10 | — | 4 2 | — | 5 15 | 7 16 | — | 9 20 | — | — | .. | .. | 8 47 | .. |
| 11 | Pyllo Hill arr. | — | — | — | 11 15 | — | 1 10 | — | 3 10 | — | 4 55 | — | — | 8 25 | — | 10 40 | 10 55 | .... | .... | — | .... |
| 11¾ | Bristol arr. | 7 40 | 8 40 | 9 30 | .. | 11 40 | — | 2 15 | .. | 4 5 | .. | 5 19 | 7 20 | .. | 9 25 | .. | .. | .. | .. | 8 50 | .. |

**CROSSING ARRANGEMENTS.**

The 6.55 a.m. ex Bristol crosses the 7.5 a.m. ex Portishead at Pill.
The 7.0 a.m. RR ex Bristol will cross 7.5 a.m. ex Portishead at Clifton Bridge.
The 7.45 a.m. " 8.0 a.m. " " Pill.
The 10.10 a.m. ex Bristol will cross the 10.20 a.m. RR ex Portishead at Pill.
The 10.50 a.m. " 11.5 a.m. ex Portishead at Clifton Bridge.
The 12.45 p.m. ex Bristol crosses the 12.15 p.m. RR ex Portishead at Clifton Bridge.
The 1.30 p.m. RR ex Bristol crosses the 1.40 p.m. from Portishead at Clifton Bridge and 1.45 p.m. Train ex Portishead at Pill, when it runs.
The 3.50 p.m. ex Bristol crosses the 4.0 p.m. RR ex Portishead at Pill, when that Train runs.
The 4.50 p.m. RR ex Bristol crosses the 4.40 p.m. ex Portishead at Clifton Bridge, when it runs.
The 6.10 p.m. RR ex Bristol crosses the 6.45 p.m. ex Portishead at Pill.
The 7.20 p.m. ex Bristol crosses the 7.30 p.m. RR ex Portishead at Pill, when it runs.
The 8.30 p.m. ex Bristol crosses the 8.45 p.m. ex Portishead at Pill.

**NOTE.—The loads of the Goods Trains not to exceed 28 Waggons.**

**The Portishead Goods Trains run through to the Great Western Yard.**

# Chapter Eleven

## Timetables and Train Working

The initial train service offered five trains each way on weekdays and two on Sundays, the journey time from Bristol varying from 45 to 30 minutes, the latter omitting Portbury. Trains terminated at the Bristol & Exeter station, Bristol. Fares: 6d. first class single to Clifton Bridge and 3s. first class return to Portishead. No third class returns were issued, so two singles were needed at 11½d. each. No third class season tickets were available, but a monthly second class ticket from Clifton Bridge to Portishead cost £1 5s. or £1 12s. Bristol to Portishead. A quarterly season cost £3 and £4 4s. respectively and annual tickets £9 and £11.

In 1871 excursion return fares were advertised by the 5.00 pm from Temple Meads Mondays to Fridays for 2s. 6d. first class and 1s. 6d. third class; on Mondays there was an additional excursion leaving Temple Meads at 10.20 am for the same fare, while on Saturdays the 3.10 pm only cost 1s. return.

The parcels rate was 4d. for 7 lb; 10d. for 56 lb and 1s. 2d. for a hundredweight.

By September 1869 the service offered eight trains daily in each direction and one on Sundays, 45 minutes were allowed for the journey of 11½ miles. Until about July 1875 goods traffic was carried by passengers trains. By 1877 most trains took 35 minutes.

A letter appeared in the *Bath & Cheltenham Gazette* for 26th June, 1872 complaining of the poor timetable. A passenger from Bath wishing to spend the weekend at Portishead would be obliged to return by the Sunday 8.30 pm from Portishead arriving at Temple Meads 20 minutes after the Bath train had left, necessitating him waiting four hours for the mail which arrived at Bath at 1.10 am. The alternative was catching the 7.00 am from Portishead on Monday morning reaching Bath about 9.30 as the express from Bristol left five minutes before the arrival of the train from Portishead.

By 1886 nine passenger trains ran each way, the fastest taking 35 minutes. One ran on Sundays.

In October 1903 11 passenger trains were run each way on weekdays, plus an extra on Thursdays, taking 35 minutes for the journey. An unusual feature of the line was the number of places which enjoyed a through service to and from the branch. Through trains ran from Bath, Cardiff, Severn Tunnel Junction, Weymouth, Newport, Portsmouth, Salisbury and Chipping Sodbury. Through trains were operated to Patchway, Cardiff, Severn Tunnel Junction, Salisbury, Bath, Westbury. Two trains ran over the branch each way on Sundays. Seven goods trains ran each way daily.

# THE BRISTOL TO PORTISHEAD BRANCH

Branch Working Timetable July 1898

Branch Working Timetable Summer 1914

13th July – 20th September, 1925

In 1914 13 passenger trains ran each way and two on Sundays. Some trains originated from Badminton, Bath, Stapleton Road, Cardiff and Swindon, some running to Severn Tunnel Junction, Cardiff, Frome and Trowbridge.

In July 1922 there were 12 trains each way, one of which was a steam railmotor and on Thursdays and Saturdays there was an extra late train Bristol-Portishead. Three trains on Sundays.

In the summer of 1925 13 trains ran each way daily with three on Sundays. Some down trains originated from Avonmouth and Trowbridge, while some up trains ran direct to Severn Tunnel Junction, Cardiff, Avonmouth and Henbury, the fastest taking 32 minutes between Bristol and Portishead. Three trains ran each way on Sundays.

On 8th July, 1929 the GWR introduced an hourly clock-face timetable with a half-hourly service at certain hours. One train left Temple Meads hourly from 11.30 am to 10.30 pm, while the other ran from Ashton Gate hourly from 11.10 am to 9.10 pm. Up trains left Portishead at 18 minutes past the hour for Ashton Gate and at 48 minutes for Temple Meads from 9.18 am to 10.18 pm. On Sundays the half-hourly service started at 2.00 pm. Most of the trains crossed at Pill or Portbury Shipyard. This intensive service continued with only slight modification until the outbreak of the Second World War. Latterly it was an hourly service, though half-hourly on Saturday afternoons, with all trips beginning or ending at Temple Meads.

The July 1938 time table showed 24 down trains plus five on Saturdays plus one one Wednesdays and Saturdays; the up service 22 trains plus five on Saturdays and one on Wednesdays and Saturdays; 13 each way ran on Sunday. Even in November 1944 the wartime service showed 12 down and 11 up trains on weekdays, the first leaving Temple Meads at 5.30 am and the last arriving Portishead at 11.34 pm. An additional two down and three up ran Saturdays-only, while the Sunday service was six each way. Certainly in the 1940s the locomotive working the last down passenger train returned with the last up freight. Similarly the engine of the first down morning freight, returned with the first up passenger train.

By January 1947 the service had increased to 15 down and 13 up trains on weekdays, with an extra two down and three up on Saturdays, with seven trains each way on Sundays.

In 1955 the service consisted of 12 down and 11 up trains, with lengthy gaps in the morning and afternoon; four extra trains were run on Saturdays. Although nominally the service was based on Temple Meads, in practice three down trains worked from Stapleton Road; from St Andrew's Road via Clifton Down; and from Avonmouth via Henbury

TIMETABLES & TRAIN WORKING 127

Passenger timetable 18th July-11th September, 1932 showing Nightingale Valley. Note the half hourly Sunday afternoon service.

respectively. The Sunday timetable showed eight trains, two of which started from Filton Junction and two from Severn Beach. In the 1950s, although nominally passenger trains were based on Temple Meads, in practice three down trains worked from Stapleton Road; St Andrew's Road via Clifton Down; and from Avonmouth via Henbury respectively. Four trips were made by ex-GWR diesel railcars the remainder being 4-coach sets hauled by a 2-6-2T.

In 1958 14 down and 15 up trains ran on weekdays plus three each way on Saturdays; five trains ran on Sundays. Five up and five down trains called at Ham Green Halt and the 1.30 pm ex-Temple Meads on

**Table 75 — BRISTOL and PORTISHEAD**

**Week Days**

| Miles | | Through Train from Stapleton Road am (2) | am | am | am | am | am | am | am | am | am (2) | E am | S am | S am | E am |
|---|---|---|---|---|---|---|---|---|---|---|---|---|---|---|---|
| — | **Bristol (Tem. Meads)** dep | 5 27 | 6 55 | 6 58 | 7 25 | 7 48 | 8 5 | 8 20 | 8 42 | 9 0 | 9 30 | 9 45 | 10 0 | 1012 | 1030 |
| 1 | Bedminster | 5 30 | 6 58 | 7 1 | 7 28 | 7 51 | 8 8 | 8 23 | 8 45 | 9 5 | 9 33 | 9 49 | 10 3 | 1016 | 1033 |
| 1½ | Parson Street | 5 34 | 7 0 | 7 4 | 7 30 | 7 54 | 8 11 | 8 25 | .. | 9 8 | 9 36 | 9 52 | 10 6 | 1019 | 1036 |
| 3 | Ashton Gate | .. | .. | 7 7 | .. | .. | 8 14 | .. | .. | .. | 9 39 | .. | 10 9 | .. | 1039 |
| 3½ | Clifton Bridge | 5 39 | .. | 7 10 | .. | .. | 8 16 | .. | .. | .. | 9 41 | .. | 1011 | .. | 1041 |
| 7 | Ham Green Halt | .. | .. | .. | .. | .. | .. | .. | .. | .. | .. | .. | .. | .. | .. |
| 7¾ | Pill | 5 48 | .. | 7 19 | .. | .. | 8 29 | .. | .. | .. | 9 50 | .. | 1020 | .. | 1050 |
| 9¼ | Portbury | .. | .. | .. | .. | .. | 8 35 | .. | .. | .. | .. | .. | 1027 | .. | 1056 |
| 11¼ | **Portishead** arr | 5 58 | .. | 7 27 | .. | .. | 8 40 | .. | .. | .. | 10 0 | .. | 1032 | .. | 11 1 |

**Week Days—continued**

| | E am | S am (2) | S am | B am | S am | E am (2) | S am (2) | E am | am | S pm | pm A | E pm | S pm | S pm | E pm | S pm |
|---|---|---|---|---|---|---|---|---|---|---|---|---|---|---|---|---|
| **Bristol (Tem. Meads)** dep | 1040 | .. | .. | 11 9 | 1120 | 1130 | .. | 1138 | 1153 | 12 8 | 1230 | 1240 | 1257 | 1 0 | 1 10 | 1 22 |
| Bedminster | 1045 | .. | .. | 1114 | 1126 | 1133 | .. | 1143 | 1157 | 1212 | 1233 | 1244 | 1 1 | 1 3 | 1 14 | 1 25 |
| Parson Street | 1049 | .. | .. | 1118 | 1130 | 1136 | .. | 1146 | .. | .. | 1236 | 1247 | 1 3 | 1 6 | 1 16 | 1 27 |
| Ashton Gate | .. | 1045 | 1115 | .. | .. | 1139 | 1145 | .. | .. | .. | 1239 | .. | .. | 1 9 | .. | .. |
| Clifton Bridge | .. | 1047 | 1117 | .. | .. | 1141 | 1147 | .. | .. | .. | 1241 | .. | .. | 1 11 | .. | .. |
| Ham Green Halt | .. | .. | .. | .. | .. | .. | .. | .. | .. | .. | 1248 | .. | .. | .. | .. | .. |
| Pill | .. | 1057 | 1127 | .. | .. | 1150 | 1156 | .. | .. | .. | 1250 | .. | .. | 1 22 | .. | .. |
| Portbury | .. | 11 3 | .. | .. | .. | .. | .. | .. | .. | .. | 1256 | .. | .. | 1 31 | .. | .. |
| **Portishead** arr | .. | 11 8 | 1137 | .. | .. | 12 0 | 12 6 | .. | .. | .. | 1 1 | .. | .. | 1 36 | .. | .. |

**Week Days—continued**

| | pm V | S pm | E pm | pm A | S pm | S pm | E pm | E pm (2) | S pm (2) | E pm | S pm | E pm | D pm (2) | pm | pm | pm | E pm |
|---|---|---|---|---|---|---|---|---|---|---|---|---|---|---|---|---|---|
| **Bristol (Tem. Meads)** dep | 1 30 | 1 55 | 2 0 | 2 30 | 2 35 | 2 55 | 3 2 | 3 30 | .. | 3 50 | 4 0 | 4 10 | 4 25 | 4 35 | 4 42 | 5 15 | 5 20 |
| Bedminster | 1 33 | 1 59 | 2 3 | 2 33 | 2 39 | 2 58 | 3 6 | 3 33 | .. | 3 53 | 4 3 | 4 15 | 4 30 | 4 38 | 4 48 | 5 18 | 5 23 |
| Parson Street | 1 36 | 2 3 | 2 6 | 2 36 | 2 44 | 3 1 | 3 9 | 3 36 | .. | 3 56 | 4 6 | 4 18 | 4 35 | 4 41 | 4 51 | 5S20 | 5 26 |
| Ashton Gate | 1 39 | .. | .. | 2 39 | .. | .. | .. | 3 39 | 3 39 | 3 59 | .. | .. | .. | 4 44 | .. | .. | .. |
| Clifton Bridge | 1 41 | .. | .. | 2 41 | .. | .. | .. | 3 41 | 3 41 | 4 2 | .. | .. | .. | 4 46 | .. | .. | .. |
| Ham Green Halt | aa | .. | .. | 2 48 | .. | .. | .. | .. | .. | .. | .. | .. | .. | 4 53 | .. | .. | .. |
| Pill | 1 50 | .. | .. | 2 50 | .. | .. | .. | 3 50 | 3 50 | 4 11 | .. | .. | .. | 4 55 | .. | .. | .. |
| Portbury | 1 57 | .. | .. | 2 56 | .. | .. | .. | .. | .. | .. | .. | .. | .. | 5 1 | .. | .. | .. |
| **Portishead** arr | 2 1 | .. | .. | 3 1 | .. | .. | .. | 4 1 | 4 1 | 4 29 | .. | .. | .. | 5 6 | .. | .. | .. |

**Week Days—continued**

| | pm | pm | E pm | E pm | pm | pm A | pm | pm | pm | pm | pm A |
|---|---|---|---|---|---|---|---|---|---|---|---|
| **Bristol (Tem. Meads)** dep | 5 30 | 5 32 | 5 42 | 6 4 | 6 10 | 6 30 | 6 33 | 7 30 | 9 15 | 9 43 | 11 0 |
| Bedminster | 5 33 | 5 36 | 5 46 | 6 7 | 6 14 | 6 33 | 6 37 | 7 33 | 9 18 | 9 48 | 11 3 |
| Parson Street | 5 36 | 5 39 | 5 48 | 6 10 | .. | 6 36 | 6 40 | 7 36 | 9 21 | 9 51 | 11 6 |
| Ashton Gate | 5 39 | .. | .. | 6 13 | .. | 6 39 | .. | 7 39 | 9 24 | .. | 11 9 |
| Clifton Bridge | 5 41 | .. | .. | 6 15 | .. | 6 41 | .. | 7 41 | 9 26 | .. | 1111 |
| Ham Green Halt | 5 48 | .. | .. | .. | .. | .. | .. | .. | .. | .. | 1118 |
| Pill | 5 50 | .. | .. | 6 24 | .. | 6 50 | .. | 7 50 | 9 35 | .. | 1122 |
| Portbury | 5 59 | .. | .. | .. | .. | 6 56 | .. | .. | .. | .. | .. |
| **Portishead** arr | 6 3 | .. | .. | 6 36 | .. | 7 1 | .. | 8 0 | 9 45 | .. | 1132 |

**Sundays**

| | am | am | am | am | am | noon | pm | pm | pm | pm | pm | pm | pm | pm | pm (2) | pm | pm | pm (2) |
|---|---|---|---|---|---|---|---|---|---|---|---|---|---|---|---|---|---|---|
| **Bristol (Tem. Meads)** dep | 9 0 | 1025 | 1030 | 1052 | 1125 | 12 0 | 1230 | 2 10 | 2 25 | 2 30 | 3 48 | 5 12 | 5 30 | 5 50 | 6 30 | 7 30 | 7 44 | 8 35 |
| Bedminster | 9 4 | 1030 | 1033 | 1055 | 1128 | 12 4 | 1233 | 2 13 | 2 28 | 2 33 | 3 54 | 5 16 | 5 33 | 5 54 | 6 33 | 7 35 | 7 48 | 8 36 |
| Parson Street | 9 7 | 1033 | 1036 | 1058 | 1132 | 12 7 | 1236 | 2 15 | 2 31 | 2 36 | 3 57 | 5 18 | 5 36 | 5 57 | 6 36 | 7 38 | 7 51 | 8 41 |
| Ashton Gate | .. | .. | 1039 | .. | .. | .. | 1239 | .. | .. | 2 39 | .. | .. | 5 39 | .. | 6 39 | 7 41 | .. | 8 44 |
| Clifton Bridge | .. | .. | 1041 | .. | .. | .. | 1241 | .. | .. | 2 41 | .. | .. | 5 41 | .. | 6 41 | 7 43 | .. | 8 46 |
| Ham Green Halt | .. | .. | .. | .. | .. | .. | 1248 | .. | .. | 2 48 | .. | .. | .. | .. | .. | .. | .. | .. |
| Pill | .. | .. | 1050 | .. | .. | .. | 1250 | .. | .. | 2 50 | .. | .. | 5 50 | .. | 6 50 | 7 52 | .. | 8 55 |
| Portbury | .. | .. | 1055 | .. | .. | .. | 1256 | .. | .. | 2 56 | .. | .. | 5 55 | .. | 6 55 | 7 57 | .. | .. |
| **Portishead** arr | .. | .. | 11 0 | .. | .. | .. | 1 2 | .. | .. | 3 1 | .. | .. | 6 0 | .. | 7 0 | 8 2 | .. | 9 4 |

**A** Second class only Mondays to Fridays. First and Second class on Saturdays.

**aa** On Wednesdays and Saturdays only, calls at 1 48 pm

**B** Except Saturdays and runs 22nd July to 30th August inclusive only

**D** Saturdays only until 31st August inclusive only

**E** Except Saturdays

For continuation of Notes, see page 245

244

Passenger timetable 17th June-15th September, 1957

Wednesdays and Saturdays. The 5.42 am ex-Temple Meads terminated at Ashton Gate and the 6.45 am started at Ashton Gate. Several trains omitted Portbury. Most trains took 31 minutes but the fastest down train was the 6.58 am which took only 29 minutes. The fastest up took 31 minutes. The 5.15 pm from Portishead composed of Southern Region coaches, became the 6.02 pm from Temple Meads-Bath which arrived at Bath Green Park at 6.44 pm and continued to Bournemouth West. Trains ran at regular intervals past the hour tending to leave Temple Meads at 30 minutes past the hour and leaving Portishead at 15 minutes past.

**Table 75**—*continued* **PORTISHEAD and BRISTOL**

**Week Days**

| Miles | | am | am (2) | | am | am | | am | am | | am | am | | am (2) | am | am | | am | am | | am | am | S am |
|---|---|---|---|---|---|---|---|---|---|---|---|---|---|---|---|---|---|---|---|---|---|---|---|
| — | **Portishead** dep | | | | | | | | | | | | | 6 44 | | | | | 7 55 | | | 8 18 | Not after 7th September |
| 2 | Portbury | | | | | | | | | | | | | 6 49 | | | | | 8 0 | | | 8 23 | |
| 3¼ | Pill | | | | | | | | | | | | | 6 54 | | | | | 8 5 | | | 8 28 | |
| 4½ | Ham Green Halt | | | | | | | | | | | | | | | | | | | | | | |
| 8 | Clifton Bridge | | | | | | | | | | | | | 7 3 | | | | | 8 14 | | | 8 39 | |
| 8½ | Ashton Gate | | | | | | | | | | 6 45 | | | 7 5 | | | | | 8 16 | | | 8 41 | |
| 9¾ | Parson Street | 5 16 | 5 50 | | | 6 25 | | 6 33 | 6 37 | | 6 49 | 6 58 | | 7 9 | 7 11 | | | 8 10 | 8 20 | | 8 31 | 8 44 | |
| 10½ | Bedminster | 5 19 | 5 54 | | 6 1 | 6 29 | | 6 36 | 6 41 | | 6 52 | 7 1 | | 7 12 | 7 14 | 7 49 | | 8 13 | 8 23 | | 8 34 | 8 47 | 8 52 |
| 11½ | **Bristol (Tem. Meads)** arr | 5 23 | 6 0 | | 6 4 | 6 32 | | 6 41 | 6 45 | | 6 55 | 7 5 | | 7 15 | 7 19 | 7 52 | | 8 16 | 8 26 | | 8 38 | 8 51 | 8 55 |

**Week Days**—*continued*

| | am | | am | S am | E am (2) | S am (2) | S am | E am | | S am (2) | S pm | S am | pm (2) | | E pm | | pm | | pm A | | E pm | | S pm |
|---|---|---|---|---|---|---|---|---|---|---|---|---|---|---|---|---|---|---|---|---|---|---|---|
| **Portishead** dep | | | 9 0 | | 1015 | 1015 | 1045 | 1115 | | 1115 | | 1145 | 1215 | | | | | | 1 15 | | | | 1 45 |
| Portbury | | | | | 1020 | 1020 | | | | | | | 1220 | | | | | | | | | | 1 50 |
| Pill | | | 9 10 | | 1025 | 1028 | 1056 | 1125 | | 1125 | | 1156 | 1225 | | | | | | 1 25 | | | | 2 0 |
| Ham Green Halt | | | 9 12 | | | | | | | | | | 1227 | | | | | | | | | | 2 3 |
| Clifton Bridge | | | 9 21 | | 1034 | 1037 | 11 5 | 1134 | | 1135 | | 12 8 | 1236 | | | | | | 1 34 | | | | 2 12 |
| Ashton Gate | | | 9 23 | | 1036 | 1039 | 11 7 | 1136 | | 1137 | | 1210 | 1238 | | | | | | 1 36 | | | | 2 14 |
| Parson Street | 9 4 | | 9 27 | | 1040 | | | 1140 | | | 12 9 | 1214 | 1242 | | 1255 | | 1 16 | | 1 40 | | 2 12 | | 2 17 |
| Bedminster | 9 7 | | 9 30 | 1032 | 1043 | | | 1143 | | | 1212 | 1217 | 1245 | | 1 0 | | 1 19 | | 1 43 | | 2 15 | | 2 20 |
| **Bristol (Tem. Meads)** arr | 9 11 | | 9 34 | 1036 | 1046 | | | 1146 | | | 1215 | 1220 | 1248 | | 1 5 | | 1 26 | | 1 46 | | 2 18 | | 2 23 |

**Week Days**—*continued*

| | E pm | S pm (2) | S pm | E pm | | pm A | S pm | E pm | | pm (2) | | | pm | | pm | | pm | | E pm | | E pm | | pm |
|---|---|---|---|---|---|---|---|---|---|---|---|---|---|---|---|---|---|---|---|---|---|---|---|
| **Portishead** dep | 2 15 | 2 15 | | | | 3 15 | | | | 4 15 | | | | | 5 15 | | | | 5 47 | | | | 6 15 |
| Portbury | 2 20 | 2 20 | | | | 3 20 | | | | | | | | | 5 20 | | | | | | | | |
| Pill | 2 25 | 2 25 | | | | 3 25 | | | | 4 25 | | | | | 5 25 | | | | 6 0 | | | | 6 25 |
| Ham Green Halt | 2 27 | 2 27 | | | | | | | | 4 28 | | | | | | | | | | | | | |
| Clifton Bridge | 2 36 | 2 36 | | | | 3 34 | | | | 4 36 | | | | | 5 34 | | | | 6 9 | | | | 6 34 |
| Ashton Gate | 2 38 | 2 38 | | | | 3 36 | | | | 4 38 | | | | | 5 36 | | | | 6 11 | | | | 6 36 |
| Parson Street | 2 42 | | 2 55 | 3 9 | | 3 40 | 3 52 | 4 31 | | 4 41 | | | 5 28 | | 5 40 | | 5 48 | | 6 15 | | 6 27 | | 6 40 |
| Bedminster | 2 45 | | 2 58 | 3 12 | | 3 43 | 3 55 | | | 4 44 | | | 5 31 | | 5 43 | | 5 51 | | 6 18 | | 6 30 | | 6 43 |
| **Bristol (Tem. Meads)** arr | 2 48 | | 3 2 | 3 16 | | 3 46 | 3 59 | 4 36 | | 4 48 | | | 5 36 | | 5 46 | | 5 56 | | 6 21 | | 6 34 | | 6 46 |

**Week Days**—*continued*

| | pm A | | P pm | pm | E pm | | pm | pm | pm | | pm | pm | pm | D pm (2) | | pm | | pm | | pm | | H night (2) | |
|---|---|---|---|---|---|---|---|---|---|---|---|---|---|---|---|---|---|---|---|---|---|---|---|
| **Portishead** dep | 7 15 | | | | | | | | 8 15 | | | | | | | | | 1015 | | | | | |
| Portbury | 7 20 | | | | | | | | | | | | | | | | | | | | | | |
| Pill | 7 25 | | | | | | | | 8 25 | | | | | | | | | 1025 | | | | | |
| Ham Green Halt | | | | | | | | | | | | | | | | | | | | | | | |
| Clifton Bridge | 7 34 | | | | | | | | 8 34 | | | | | | | | | 1034 | | | | | |
| Ashton Gate | 7 36 | | | | | | | | 8 36 | | | | | | | | | 1036 | | | | | |
| Parson Street | 7 40 | | 7 50 | 7 57 | 8 4 | | 8 19 | 8 32 | 8 40 | | 9 3 | 9 10 | 9 38 | 10 5 | | 1018 | | 1040 | | 1110 | | 1242 | |
| Bedminster | 7 43 | | 7 54 | 8 0 | 8 7 | | 8 22 | 8 35 | 8 43 | | 9 6 | 9 14 | 9 42 | 1010 | | 1021 | | 1043 | | 1113 | | | |
| **Bristol (Tem. Meads)** arr | 7 46 | | 8 1 | 8 5 | 8 11 | | 8 28 | 8 40 | 8 46 | | 9 10 | 9 21 | 9 51 | 1017 | | 1028 | | 1046 | | 1116 | | 1248 | |

**Sundays**

| | am (2) | am | | am | am | am | | pm | pm | | pm | pm | pm (2) | pm | | pm | pm | pm | pm | pm (2) | | pm | pm |
|---|---|---|---|---|---|---|---|---|---|---|---|---|---|---|---|---|---|---|---|---|---|---|---|
| **Portishead** dep | | | | | | 1115 | | 1 15 | 4 40 | | | 6 15 | 7 15 | | | | | 8 15 | | 9 30 | | | |
| Portbury | | | | | | 1120 | | 1 20 | 4 45 | | | 6 20 | 7 20 | | | | | 8 20 | | | | | |
| Pill | | | | | | 1125 | | 1 25 | 4 50 | | | 6 25 | 7 25 | | | | | 8 25 | | 9 40 | | | |
| Ham Green Halt | | | | | | | | | 4 53 | | | 6 28 | | | | | | | | | | | |
| Clifton Bridge | | | | | | 1134 | | 1 34 | 5 0 | | | 6 35 | 7 34 | | | | | 8 34 | | 9 49 | | | |
| Ashton Gate | | | | | | 1136 | | 1 36 | 5 3 | | | 6 38 | 7 36 | | | | | 8 37 | | 9 52 | | | |
| Parson Street | 1242 | 6 37 | | 8 39 | 1033 | 1140 | | 1 40 | 5 7 | | 5 57 | 6 42 | 7 40 | 8 11 | | 8 18 | 8 38 | 8 41 | 9 13 | 9 56 | | 10 0 | 1017 |
| Bedminster | | 6 41 | | 8 43 | 1036 | 1143 | | 1 43 | 5 10 | | 6 0 | 6 45 | 7 43 | 8 15 | | 8 21 | 8 41 | 8 44 | 9 16 | 9 59 | | 10 3 | 1020 |
| **Bristol (Tem. Meads)** arr | 1248 | 6 45 | | 8 47 | 1040 | 1147 | | 1 47 | 5 13 | | 6 4 | 6 48 | 7 47 | 8 20 | | 8 25 | 8 45 | 8 47 | 9 20 | 10 2 | | 10 6 | 1024 |

H Saturday nights only
P Except Saturdays. Runs 1st July to 30th August inclusive only

S or S Saturdays only
V Second class only on Saturdays

(2) Second class only

For other notes, see page 244

245

Passenger timetable 17th June-15th September, 1957

In March 1962 the 14 trains daily were reduced to just six peak hour services each way and the five each way Sunday services were withdrawn completely.

The penultimate timetable 9th September, 1963 – 4th June, 1964 showed six down trains each weekday, with one less on Saturdays and 6 up trains. One each way was marked 'Second class only' – indicating diesel working. No down train was operated between 8.15 am and 4.37 pm

The final timetable showed six trains each way on weekdays, with two extra down and one up on Saturdays. The 'Second class only', 16.35 Mondays-Fridays from Portishead, actually worked through to Bath, Green Park where it arrived at 17.49. Trains on the Portishead branch were timed to take about 30 minutes between Bristol Temple Meads and Portishead, or vice versa.

In August 1964 six trains each way daily, but trains left at poor times with nothing from Temple Meads to Portishead between 08.30 and 14.30, or from Portishead for Temple Meads between 09.15 and 17.15. The 3s. return rail fare Portishead -Temple Meads was 1s. less than the bus fare and offered 28 minutes running time compared with 55 minutes on the bus. The branch enjoyed 450 rail passengers per week.

The goods timetable for the summer of 1925 showed five each way between Bristol and Portishead. No goods trains were run on Sundays.

The goods timetable in 1955 comprised the 5.00 am and 8.15 am down from West Depot to Portishead. Up workings were the 12.20 am to Ashton Meadows, the 3.20 pm to Stoke Gifford and the 8.12 pm to Ashton Meadows, 0-6-0PTs handling these workings. Timing for the run between West Depot and Portishead varied from 35 minutes non-stop to 1 hour 31 minutes for a pick-up freight. Timber and oil traffic were particularly heavy.

Table 47                                                                                         Weekdays on

**Bristol to Portishead** | **Portishead to Bristol**

| | | B 50 | C | B 50 | C | A | | | | | | | | |
|---|---|---|---|---|---|---|---|---|---|---|---|---|---|---|
| BRISTOL TEMPLE MEADS .. d | 5 27 | 6 58 | 8(25 | 8 30 | 14(30 | 16(35 | 17 47 | 18(45 | PORTISHEAD .. .. .. d | 6 30 | 7 53 | 8 15 | 9(15 | 17 15 | 18 25 .. 19 |
| BEDMINSTER.................. d | 5 32 | 7 02 | 8(28 | 8 34 | 14(33 | 16(39 | 17 51 | 18(48 | PILL ........................... d | 6 39 | 8 03 | 8 25 | 9(25 | 17 25 | 18 35 .. 19 |
| PARSON STREET .. .. .. d | 5 34 | 7 04 | 8(31 | 8 37 | 14(36 | 16(42 | 17 55 | 18(51 | HAM GREEN HALT .. .. .. | .. | .. | .. | 9(27 | .. | .. |
| ASHTON GATE HALT .......... d | .... | 7 08 | 8(34 | 8 40 | 14(39 | 16(45 | 17 58 | 18(54 | CLIFTON BRIDGE HALT .......... d | 6 48 | 8 12 | 8 34 | 9(36 | 17 34 | 18 44 .. 19 |
| CLIFTON BRIDGE HALT .. .. d | 5 39 | 7 10 | 8(36 | 8 42 | 14(41 | 16(47 | 18 00 | 18(56 | ASHTON GATE HALT .. .. .. | 6 50 | 8 14 | 8 36 | 9(38 | 17 36 | 18 46 .. 19 |
| HAM GREEN HALT .. .. .. d | .... | .... | .... | .... | 14(48 | 16(54 | 18 07 | .... | PARSON STREET ............... | 6 54 | 8 18 | 8 41 | 9(42 | 17 40 | 18 50 .. 19 |
| PILL .......................... d | 5 48 | 7 19 | 8(46 | 8 51 | 14(50 | 16(56 | 18 09 | 19(05 | BEDMINSTER .................. | 6 57 | 8 21 | 8 44 | 9(45 | 17 43 | 18 53 .. 19 |
| PORTISHEAD................ a | 5 58 | 7 28 | 8(55 | 9 00 | 15(01 | 17(06 | 18 19 | 19(16 | BRISTOL TEMPLE MEADS .... a | 7 02 | 8 24 | 8 49 | 9(49 | 17 46 | 18 56 ... 20 |

Heavy figures indicate through carriages   A  Not Saturdays from 12 September
For general notes see page 49                B  Until 5 September
                                             C  Not Saturdays 20 June to 5 September

The final passenger timetable June 1964. Note the eight-hour gap in services on Mondays to Fridays.

# Chapter Twelve

## Signalling and Permanent Way

The line was converted to standard gauge between Saturday 24th January, 1880 and Tuesday 27th January, 1880. The change of gauge meant a re-inspection of the line by the Board of Trade which would have insisted on interlocking, this was carried out by Walter Easterbrook in 1879-1880. The line was doubled from Portishead to Clifton Bridge on 2nd September, 1883.

In 1886 the section Clifton Bridge-Pill used a white triangular staff and ticket, while Pill-Portishead had a round blue staff and ticket. In 1896 the train staff and ticket sections were replaced by the electric train staff. Due to extreme weather in the winter of 1940 bringing down telegraph wires, the block system had to be abandoned and time interval working instituted.

A new signal box at Portishead opened on 4th January, 1954 controlling track-circuited lines as far as the Down distant, set at 1,923 yards from the box. Signals were upper quadrant except for the colour light platform starting signals. The upper quadrants were rare examples of a Western Region pattern and the to move up the post to push the arm 'off'. This was because they were lower quadrant parts inverted. The box had an indoor WC – unusual at that date when older boxes normally had an outside facility.

BR had made a national ruling that lower quadrant signals were only to be provided in existing lower quadrant areas. Regarding the new station at Portishead, the signal sighting committee had made their decisions with the expectation of lower quadrant signals, but upper quadrants were supplied. In the case of the signals governing exit from the two platform roads, the arm was only visible in the 'on' position, hidden behind the platform canopies when cleared, so colour lights were substituted.

With the modernization at Portishead, the remaining electric staff apparatus on the branch was withdrawn and all sections were worked by electric token.

Regulations stated that in the event of a failure of the through token instruments between Portishead and Clifton Bridge necessitating pilot working being introduced, the station masters at Clifton Bridge, Pill and Portishead must be immediately informed and confer on the telephone as to the best method of dealing with the situation. Unless a motor vehicle was immediately available to convey a pilotman from Portishead to Clifton Bridge, or vice versa, they were to arrange for two pilotmen to be appointed to work between Clifton Bridge and Pill and Pill and

Portishead respectively, the working being inaugurated from either end of the section according to the direction from which the next train was to proceed.

Just before closure to passengers Pill and Portbury Shipyard signal boxes closed on 14th April, 1964. The new token section was Clifton Bridge to Portishead. Portishead signal box closed on 5th April 1965, the method of working then became 'one engine in steam' using a wooden square train staff coloured blue from Clifton Bridge to Portishead. On 4th November, 1966 Clifton Bridge box closed and the 'one engine in steam' section was extended back to Ashton Jn, this being converted to train staff and ticket working on 31st January, 1972.

On 1st January, 1951 the Portishead branch had three permanent way gangs: seven men at Clifton Bridge; five at Pill while Portishead had a strength of seven men and a roadman, the latter dealing with tarmac surfaces. On the same date the Bristol Harbour and Wapping Wharf line was maintained by a gang of eight permanent way men plus two roadmen, while the Canon's Marsh branch had seven plus two.

With the opening to Portbury Dock the first part of the route from Parson Street had a 775 metre passing loop with colour light signalling controlled from the Bristol area signalling centre. At Ashton Junction the level crossing was monitored by closed circuit TV. The remainder of the line was operated as single line block section with stop boards and no-signalman token, so consecutive trains could run in the same direction. Within the port there is CCTV control of train movements.

Oak Wood signal box 10th April, 1953. On the left is a setting-down post for receiving the single line token, the post for picking up the new token being close to the box.

*Dr. A. J. G. Dickens*

# Chapter Thirteen

# Regeneration

With the opening of the Royal Portbury Dock on 8th August, 1978, capable of accommodating larger vessels than Avonmouth, in due course, rail access was laid at a cost of £21m. The existing moth-balled line from Parson Street Junction to Pill was to be upgraded, but Pill Junction-Portishead continued to be mothballed except for the very end which Hydrex used to test its road-rail machinery. In 1998 Railtrack commissioned an in-depth study on the benefit of re-opening the line to Portishead. Most of the Pill residents were against the re-opening of the line due to their fear of noise from trains.

The first practical sign of re-opening was on 25th April, 1999 when class '37' No. 37716 arrived on the branch with the first load of continuous welded rail which was dropped between Parson Street Junction and Ashton Junction. This was the first movement over the branch since November 1991 when the final engineer's train left the then closed Bower Ashton chief civil engineer's depot.

Principal contractors for bringing the track up to standard was Alfred McAlpine Civil Engineering. In January 2001 one of the first acts was to run a rail-mounted flail along the line which had become overgrown and enclosed by a canopy of vegetation. Coppicing cleared overhanging trees, but rare plants and trees were safeguarded or

Remains of Pill station, 16th October, 2001.  *Author*

View from the east end of Pill viaduct towards the tunnel, 29th June, 2001. The old track is in place, some of the growth has been cleared and new fencing erected.   *Author*

transplanted. Tunnels were checked for bats and checks made for badger setts. As the gorge section had no road access but could be reached from a nearby cycleway, project personnel were provided with a fleet of bicycles.

Recovery of material was begun at the Pill end and taken by rail to Ashton Gate where rails and sleepers were sent for scrap or recycling. When this work was complete track laying commenced at the Bristol end using 600 foot lengths of rail on steel sleepers. Between February and September 2001 ballast arrived from Westbury and as the track layout at Parson Street made it difficult to handle down trains, the ballast wagons arrived on the up line via a reversal at Taunton. Some works trains were 'top and tailed' (an engine at either end), while others just propelled were headed by a converted brake van with warning equipment. Empty ballast trains returned to Westbury via Bath.

Fencing and weakened structures were repaired – such as some of the bricks in Pill viaduct which were flaking. To strengthen the viaduct, liquid cement grout was placed behind the arch brickwork. In the week-ending 15th December, 2001 a heavy train was run back and forth across the viaduct and the deflections measured. The track relaying was completed on 14th August, 2001 and track for the 1.2 mile-long spur over the embankment from Pill to Portbury Dock was laid in September.

Flaking brickwork on the south side of Pill viaduct, 29th June, 2001. *Author*

The western portal of Pill tunnel; part of the old track is *in situ*, but has been lifted before reaching the tunnel, 29th June, 2001. *Author*

The inaugural train to Royal Portbury Dock on 21st December, 2001 hauled by former PBA 0-6-0ST *Portbury* approaching Pill viaduct.  *Author*

At the Royal Portbury Dock a fork lift truck places a ramp into position against wagons in order that vehicles may be loaded, 9th September, 2014.  *Author*

The line was made suitable for W10w loading gauge allowing 9 ft 6 in. containers on low platform wagons. A 100 metres of Pill Tunnel had a secondary lining which reduced clearance so the floor level was lowered and as the rock through which the tunnel was drilled was susceptible to degradation by moisture, a new drainage system was installed and the trackbed waterproofed. Within the dock perimeter the single line divides: the coal terminal to the left and the general cargo area ahead with two tracks.

A joint venture of the Bristol Port Company and National Power was an £80 million bulk handling terminal with two massive continuous-ship-unloader cranes. Initially a coal conveyor carried coal beneath the River Avon to Avonmouth where there were storage and train loading facilities, but following this installation imported coal can be handled on the Portbury side too.

The former Port of Bristol Authority 0-6-0ST *Portbury* hauled a test train over the length of line to the dock on 20th December, 2001 and then the following day, 21st December, hauled the first official train consisting of three passenger coaches containing among others, Stephen Byers MP, Minister of Transport, which left Parson Street at 11.00 am, passed Pill at 11.25 at 28-30 mph and was due at Portbury at 11.45. A rail tour ran to Pill Junction on 29th December, 2001, while on 30th December English, Welsh & Scottish Railway Ltd (EWS) carried out a route training run using No. 56083. The first revenue earning services started on 7th January, 2002 when two EWS-powered coal trains ran from Portbury Dock to Filfoots Point, Newport, No. 66250 hauling the first. From 14th January, 2002 traffic increased to seven trains daily to Filfoots, Newport and various power stations in the Midlands. A few days after opening, the rear four-wheeled wagon of one of these merry-go-round trains was derailed on the approach to Temple Meads, but jumped back on the track. Platforms 3 & 4 were necessarily closed for several hours while safety checks were made.

Residents of Pill living near the line and perhaps purchasing their property when the line was 'moth-balled', complained of noise made by trains at night as, for example, on 5th March, 2002 when trains passed at midnight, 5.30 am and 7.10 am. Conditions were ameliorated in the summer when quieter-running coal wagons were used. Royal Portbury Dock was the largest car importing terminal in the UK, about half a million cars arriving annually. On 11th September, 2002 Freightliner Heavy Haul commenced running a train from Portbury-Mossend (Glasgow) for Autologic, each carrying about 200 vehicles. At Portbury vehicles are loaded on to car-carrying wagons where a ramp is forklifted

An HST runs an excursion to the Royal Portbury Dock in 2009. *Author's collection*

to one end of the train and pushed tightly against wagons; and as clearance is very tight, a man at the top of ramp indicates with hand signals to a driver if he should go to right or left.

On 9th June, 2002 the first diesel-hauled passenger train ran actually to within dock property, an enthusiasts' excursion originating from Crewe. On 13th/14th September, 2003 the *Evening Post* and the *Western Daily Press* organized steam-hauled trips from Temple Meads to Portbury Dock hauled by BR Standard class '4' 2-6-0 No. 76079, while in 2009 the line was traversed fairly slowly by a High Speed Train.

On 8th March, 2002 villagers at Pill were highly concerned at seeing two trains come face-to-face. One had travelled from the dock loaded with coal and stopped near the M5 Avon Bridge, the other was empty and heading towards it and stopped about 50 yards from the other.

An official of the English, Welsh & Scottish Railway explained that it was a safe procedure. A train had left the terminal and halted at the exit stop board , but because he had completed his allotted hours, he could no longer remain at the controls and had to end his shift. The driver of the empty train arriving from Pill, held the token which ensured there was only one train running on the branch. As he approached the halted train he slowed and stopped. He was then asked by the shunter to reverse the loaded train back into a siding. This allowed him to return

# REGENERATION 139

The first rail tour to the Royal Portbury Dock is hauled by an EWS Class '37'.
*Author's collection*

Uncoupling EWS No. 66078 at the Royal Portbury Dock from a recently arrived train in order that it may run round, 9th September, 2014. The wagon is branded SNCF so has originated from France. *Author*

to his own locomotive and drive it to the coal terminal for the wagons to be loaded.

In 2004 about four trains ran to Portbury Dock during the day and five at night. During the rush hour, due to the amount of traffic it was particularly difficult to lower the barriers at Ashton Gate. In July 2007 Freightliner Heavy Haul won a five year contract for carrying coal from Portbury to Rugeley power station. In 2008 North Somerset Council purchased the trackbed between Portbury Junction and Portishead to safeguard it. On 2nd April, 2009 six trains ran to Rugeley, but this had fallen to two daily in late April. In 2012 two daily coal trains operated to Ferrybridge; a car train to Warrington and a Biomass train to Drax.

1st March, 2013 contractors started clearing vegetation on the part of the branch owned by the North Somerset Council in anticipation of the line re-opening to passenger traffic in 2017. On 2nd March, 2013 No. 66077 worked the first loaded gypsum train from Portbury, but this ceased on 25th January, 2017 and at present traffic is cars, wood chippings and containers. On 11th September, 2018 Rail Operations Group No. 47812 departed from Portishead with the latest Trans Pennine Express Mk 5 carriages hauling them to Longsight. A week later the same locomotive was attached to the fifth coaching set to arrive at Portbury. After waiting two hours No. 47812 was uncoupled and departed light engine. It was believed that there were issues with stock on the branch having received damaged by overhanging vegetation.

Pill: view east from Station Bridge 16th October, 2001; new track has been laid with central check rails to prevent derailed vehicles from plunging off the viaduct. *Author*

## Chapter Fourteen

## Proposed Re-opening to Passenger Traffic

On 10th May, 1989 a Bill received Royal Assent for constructing a light railway from Wapping Road via Portbury to Portishead. Costing an estimated £28.5m, it was said to be 'cheap' as it was to make use of the redundant BR route from Wapping Wharf-Portishead. Opening was expected in 1992 with a 10-minute frequency and a 17 minute journey including eight intermediate stops. The electrically-powered vehicles were planned to have a capacity of 200 passengers, thus replacing more than 160 cars. Avon County Council favoured the scheme, but its future was in doubt when Bristol County Council failed to give it support so lack of funds prevented the work from proceeding.

Despite the horrendous traffic situation between Portishead and Bristol at rush hours, the Bristol Metro has yet to get off the ground so Portishead is one of the largest towns in England today without a railway station. Portishead Railway Group was established in 2000 to lobby MPs and be represented at meetings with other campaign groups and interested parties. A study in 2010 showed that travel time from Portishead would be 17 minutes by rail compared to an hour by road during peak times.

In April 2015 local councillors decided that the new station at Portishead would be built on a corner site between Harbour Road and Quays Avenue and thus will not require the provision of a level crossing as the Office of Rail Regulator would not permit a level crossing in any plans for the new station site. In March 2016 it was announced that the opening to Portishead would be delayed until at least early 2019 due to construction problems – the track needed upgrading to operate passenger trains at 50 mph and the track geometry had to be adjusted to straighten a series of curves through the gorge, double track laid through Pill, and signalling issues dealt with.

In March 2017 MetroWest announced that the estimated costs had increased to £175m and the target opening date pushed back to 2021. Costs had spiralled because as most of it was single track, trains would need to travel at 50 mph plus to make a half-hourly service possible. To reduce the expense, it was planned to reopen with an hourly service and then make improvements to allow half-hourly trains.

Re-opening the disused section between Pill Junction and Portishead required planning consent through a Development Consent Order, but this was not needed for the remainder of the scheme as it was already within an existing operational railway. The business case showed the scheme's 'value for money' providing £3 of benefits for every £1 invested. The present timescale for the re-opening to passenger service is: autumn 2019 decision by the Secretary of State; winter 2019 signing off the planning conditions and mobilizing the contractors, while in the spring of 2020 construction would commence on the principal works and is expected to take about 18 months.

Should MetroWest Phase One come about, it will increase the UK national network for passenger trains by 14 km; bring an additional 50,000 people within the immediate catchment of the rail network; run through environmentally-important areas of the Avon Gorge and a Special Area of Conservation; enable passengers to travel from Portishead to Temple Meads in about 23 minutes; have a line speed of 75 mph between Portishead and Pill and 30 mph from Pill to Bower Ashton. The 3-car dmus will have up to 270 seats.

1.6km of existing single track will need upgrading to double track between Bower Ashton and Ashton Gate level crossing, while an intermediate signal is required on the single line in the Avon Gorge about midway. Parson Street Junction needs upgrading, as although the existing freight line from Portbury has a section of double track approaching Parson Street Junction, it has only a single track connection with the main line and this will need to be doubled to provide sufficient capacity for carrying both freight and passenger trains.

Re-opening is not as simple as renewing the track between Pill Junction and Portishead and building a new terminal station, there are a multitude of financial, legal, engineering, environmental, technical and risk issues that have to be dealt with. The preliminary case for Phase 1 of MetroWest scheme was approved by the West of England Partnership in September 2014 confirming the project was good value for money. In July 2018 it was stated the the reopening project was still 'firmly on track' claiming the chance of it receiving the £116 million needed for completion as 'reasonable'.

# Chapter Fifteen

## The Bristol Harbour Railway & Wapping Wharf Branch

---

### BRISTOL HARBOUR RAILWAY.

The public are informed that the GREAT WESTERN and BRISTOL and EXETER RAILWAY COMPANIES are now prepared to convey TRAFFIC to or from the WHARF DEPOT near PRINCE's STREET BRIDGE (thereby affording convenient and speedy access to the FLOATING HARBOUR and DOCKS).

For particulars as to Rates, &c., apply to
    Mr. HEARNE, Great Western Goods' Station, Bristol;
or at either of the undermentioned Offices :—
    ASSEMBLY ROOMS, Prince Street
    UNIVERSAL OFFICE, High Street;
    QUEEN's HEAD OFFICE, Redcliff Street.

N.B.—Until further notice no SINGLE Piece of TIMBER, MACHINERY, STONE, or any other Article exceeding TWO TONS in weight will be carried.

BY ORDER

---

Notice re Bristol Harbour Railway from the Bristol & Exeter timetable, 1877.

The Portishead branch had two subsidiary branches: the Bristol Harbour Railway and to Canon's Marsh.

When the railway first came to Bristol it did not serve the docks these being inaccessible without expensive demolition of property. In due course it was found essential to provide a link and this was the Bristol Harbour Railway, incorporated 28th June, 1866, owned jointly by the Great Western, the Bristol & Exeter and Bristol Corporation. It was an expensive three-quarters of a mile line requiring the demolition of a vicarage, the making of a cutting and a 292-yard long tunnel through a burial ground involving the expense of removal and re-interment at Arnos Vale on a plot of land purchased by the railway. The line also required the construction of a long viaduct and three bridges.

The most interesting work was an opening bascule bridge over Bathurst Basin designed by Charles Richardson, who was to become chief engineer to the Severn Tunnel and had invented the spring-handle cricket bat. Mixed gauge track was laid on the bridge decking as both gauges were in use in the city. It was moved by a horizontal steam engine built by the Avonside Engine Company in 1872 now preserved . The 250-ton bridge was delicately balanced and in order to prevent it being overdriven, was given a friction drive to the gear wheels which actually moved the bridge. Steam was supplied from a railmotor boiler.

Mrs Hare, Mayoress of Bristol, laying the first rail of the tramway to the Floating Harbour, 8th October, 1863.  *Illustrated London News*

Plan of the Wapping Wharf branch showing curvature and gradients.

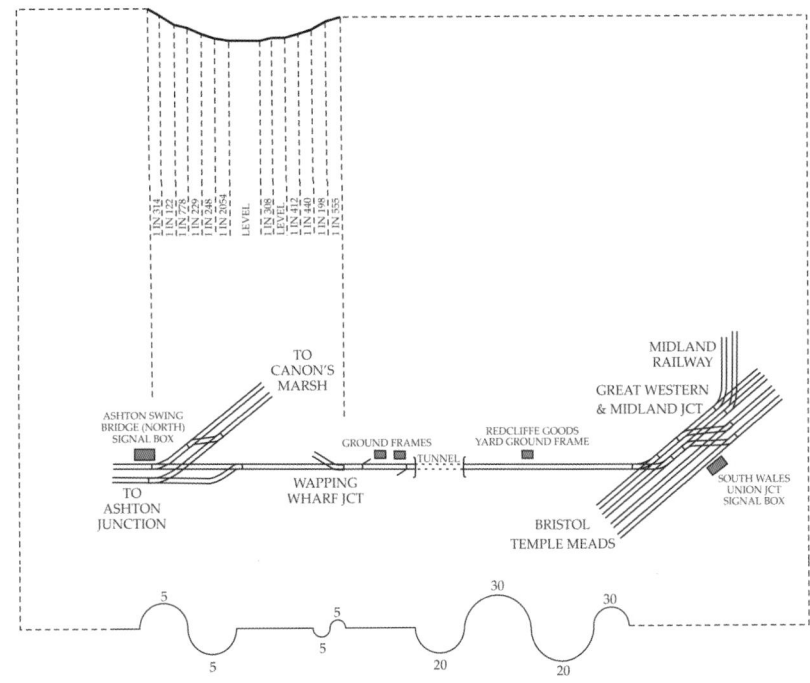

One evening in 1953, after the bridge had been raised for a ship to pass, it refused to close properly to road level leaving a gap of approximately 6 inches. No one could remember experiencing this before.

It was dark making it more difficult to locate the problem and soon there was queue of irate drivers and pedestrians impatient to reach the other side and get home.

It was a case of 'third time lucky'. A search after the third raising revealed a little tragedy. A ginger tomcat had been taking an evening stroll by the waterside. Fancying a short cut, Ginger had crossed the spot at which the bridge, when closed, rested on timber supports. It was a fatal misadventure because when the descending decking settled on him, poor Ginger was the obstruction which was preventing the bridge from closing properly.

The 65 chains of mixed gauge track from the east end of Bristol station was opened on 11th March, 1872 and extended 39 chains to Wapping Wharf on 12th June, 1876. Although a goods-only line, it was inspected by the Board of Trade officer, Colonel Yolland on 26th February, 1872. The broad gauge line had a width at formation level of 27 feet for the double track on the viaduct; on the embankment it was 30 feet and 27½ feet in the tunnel. In the two cuttings measurements were 29½ and 35 feet respectively.

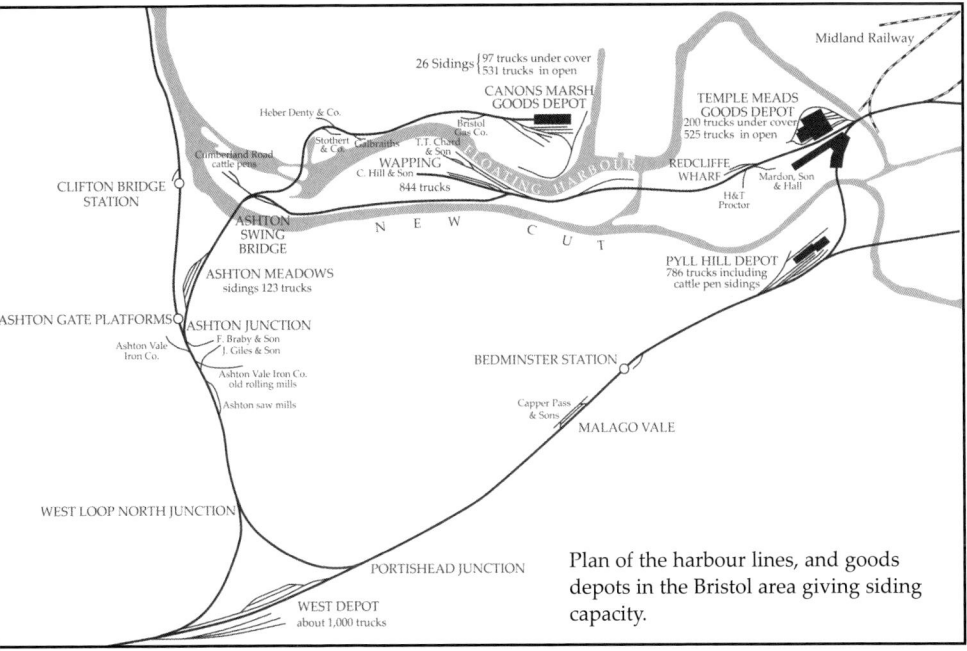

Plan of the harbour lines, and goods depots in the Bristol area giving siding capacity.

The track consisted of flat-bottomed Siemens steel laid on creosoted Baltic Redwood sleepers, the rails being secured by four ¾-inch fang bolts, sleepers next to the joints given six bolts. The 11-inch deep ballast was of broken stone at the bottom topped with furnace ash. The sharpest curve had a radius of 15 chains and the arched viaduct was 346 yards in length. Yolland suggested that a catch points be placed near the summit of the 1 in 100 incline to prevent vehicles running away over Guinea Street level crossing and that the steam supply should be automatically cut off when the bascule bridge at Bathurst Basin was nearly vertical or horizontal. The water main supplying the goods yard ran across the bridge so this required to be disconnected before the bridge was moved.

Yolland said that as the line was to be worked on the absolute block system the electrical contact would be broken when the bridge was open and this would add to safety. He said that the up line on the western side of the bridge should have catch points leading to a blind siding when the up signal was at danger.

To offer more flexibility, by an Act of 6th August, 1897 the Bristol Harbour Extension Railway was authorized to open a line from the Wapping Wharf branch 1 mile 26 chains to Ashton Junction with the Portishead branch. It opened on 4th October, 1906.

Normally a goods-only line, when the line to Weston-super-Mare was blocked between Temple Meads and Parson Street on Sundays 26th April, and 15th November, 1931 for two overline bridges to be demolished in order to further the work of track quadrupling, passenger trains were diverted over this Harbour line. Similarly when the main line was bombed during the Second World War, the Harbour line was used as a diversion. Following the demise of city docks the line fell out of use and closed east of Wapping Wharf on 6th January, 1964.

On 15th July, 1981 at 8.40 am as the Western Fuel's locomotive was hauling 35 empty HTV wagons (vacuum-braked coal hopper wagons with capacity of 21½ – 25 tons) and a brake van from Wapping Wharf to Ashton Meadows, the 35th wagon became derailed on the Ashton side of Cumberland Road crossing and dragged it to Ashton Bridge level crossing damaging the track. The derailment was caused by a slip of the retaining wall supporting the railway beside the waterway known as The Cut.

Initially train movements were restricted to within an hour either side of high tide when the water would assist in holding up the wall. The coal company then complained that this restriction of train movements was adding to its working costs. BR excavated 1.2 metres below the track and then backfilled it with spent ballast, compacted in 300mm layers with a

*Left*: A '53XX' class 2-6-0 carrying express passenger headlights, uses the Bristol Harbour line as a diversion in 1931.   *S. Miles Davey*

*Centre*: A '43XX' class 2-6-0 hauls a passenger train near Ashton Swing Bridge having used the Bristol Harbour line as a diversion in 1931/2. The photographer's 10/15 hp Fiat car HU 1837, first registered in September 1926, was parked by the road on 10th December, 1937.
*S. Miles Davey*

*Bottom*: In 1931 a diverted passenger train is about to cross Ashton Swing Bridge headed by a '4575' class 2-6-2T. Note the roadway over railway on the swing bridge.   *S. Miles Davey*

The Avon Crescent slip, 1st August, 1981, view towards Wapping Wharf.
*Author's collection*

geotextile sheet laid every 300mm. The estimated cost of this work was £15,000.

As part of the Great Western Railway 150th Anniversary Celebrations in 1985 excursions from Temple Meads to Wapping Wharf via Parson Street were sponsored by the Imperial Tobacco Company. The diesel multiple-units were painted in the appropriate GWR chocolate and cream livery, while steam workings were headed by LMS 2-6-0 No. 6443 from the Severn Valley Railway. Similarly the *Bristol Evening Post* sponsored dmu and steam trips from Temple Meads, platform 13, to Portishead.

Until May 1987 Wapping Wharf remained open for a daily coal train from Ashton Gate Junction to the Western Fuel Company's yard. It was unusual in that this BR branch was worked by a privately-owned locomotive (from 1976). The operation of the remaining lines of the Bristol Harbour Railway was taken over by the Bristol Industrial Museum on 16th April, 1995.

Regular engines working over the harbour lines were 0-6-0PTs of the '16XX', '2021', '57XX', and '94XX' classes. The '16XX' class was a modern version of the '2021' class and the '94XX' a modern version of the '57XX' class. Anything up to a 'Castle' class was allowed into Wapping Yard, 'Halls' certainly appearing regularly on meat trains to Acton.

In 1943-1944 engines regularly used at Wapping Wharf were: Nos. 1538, 3632, 4603, 4653, 7719, 7793, 8793 while those at Redcliffe were: Nos. 3632, 7728, 8714.

## THE BRISTOL HARBOUR RAILWAY & WAPPING WHARF BRANCH  149

Western Fuel Co.'s 0-6-0ST *Henbury* works a train of coal hoppers to Wapping Wharf, 16th October, 1981.  *Author's collection*

Ex-Port of Bristol Authority's Peckett 0-6-0ST *Portbury* Works No. 1764 of 1917, purchased from the Admiralty after December 1919 when it was no longer required at Portbury Shipyard. It is seen here working a passenger train on the preserved Bristol Harbour Railway, 2nd August, 1988.  *Revd Alan Newman*

1935 aerial view of the line seen on the left from Temple Meads to Redcliffe Goods Yard. Temple Meads Goods Depot is in the centre of the picture. *Author's collection*

# Chapter Sixteen

## Description of the Bristol Harbour Railway

From a line running between Brunel's passenger terminus at Temple Meads and the goods shed it crossed the 316 yard-long Redcliffe viaduct and crossed Victoria Street by a three-span bridge. The centre one, 43 feet in length spanned the street and was carried on cast-iron columns adjacent to the kerbs and two of 11 feet spanned the footwalks. This structure of riveted wrought-iron plate girders, cross girders and stringers had a timber deck and was built to carry broad gauge tracks and for some time used for mixed gauge.

'57XX' class 0-6-0PT crosses Victoria Street bridge en route to Redcliffe Goods Yard in the summer of 1960.
*Michael Jenkins*

Renewing Victoria Street Bridge, 4th December, 1960; it received little use as the line closed 6th January, 1964.
*BR*

The new Victoria Street Bridge 26th May, 1961.  *BR*

View from St Mary's Redcliffe to Temple Meads *circa* 1872.  *Author's collection*

In 1961 it was replaced and provided for future road widening allowing for two 30-foot carriageways, though initially only one was fixed. Rail level was raised 1 foot 4 inches and headroom below the bridge increased from 15 foot 8 inches to 16 foot 6 inches; each of the two spans was 55 foot 6 inches. Portal frame construction was chosen to meet the wishes of the city engineer who desired a welded steel bridge with smooth clean lines unimpaired by bolt heads and stiffeners. Four welded plate-girder portals were erected, decking of waterproof welded steel and ballasted cross-sleepered permanent way.

Redcliffe sidings ground frame (118 miles 48 chains) controlled the entrance to Redcliffe Goods Yard, in the shadow of St Mary's, closed 1st June, 1962. It was a shunting line only and thus all operations were carried on by hand signals under shunters' directions.

Redcliffe Yard dealt with a heavy coal traffic and other sidings were used by such firms as the Imperial Tobacco Company, Mardon Son &

Wagon label for traffic from Badminton to Redcliffe Siding

A Railway Correspondence & Travel Society (RCTS) special heads past Redcliffe Goods Yard headed by Class '2' 2-6-2T No. 41202 and No. 41203 28th April, 1957.

*Dr A. J. G. Dickens*

Hall and Messrs H. & T. Proctor. The yard had a capacity of 47 wagons. Gradients on the Bristol Harbour line were generally easy, but there was a short rise of 1 in 85 in the vicinity of Redcliffe Sidings.

The load of a train from Temple Meads to Wapping Wharf was limited to 60 wagons. Before leaving Redcliffe Sidings the guard or shunter in charge of a train was required, after consulting the driver, to put down a sufficient number of brakes to ensure the train being brought to a stand at Bathurst Bridge home signal. A guard or shunter was required always to ride on or in the last vehicle of every train, or the nearest suitable vehicle, and a red tail lamp fixed on the last vehicle.

Mardon's Siding ground frame, 118 miles 55 chains gave access to the Imperial Tobacco Company's siding, and also that of Mardon, Son & Hall Limited and H. & T. Proctor. The length from Redcliffe Sidings to Bathurst Bridge was single line controlled by electric token and passed through the 292 yard-long Redcliffe tunnel below the church and churchyard. Dr Gardiner was surgeon for the Harbour Railway and his name is carved halfway through the tunnel- an official carving, not graffiti. He also oversaw the removal of the bodies from the churchyard.

The tunnel was made by the cut and cover method and some of the Georgian houses in the centre of Colston Parade had to be demolished and later replaced when the tunnel was completed. One of the houses demolished was where Samuel Plimsoll was born. The tunnel set on a curve of 18 chains radius, was given a brick lining, it was dry and the ventilation good.

The RCTS special approaches the 292 yard-long Redcliffe tunnel, 26th April, 1957.
*Dr A. J. G. Dickens*

# DESCRIPTION OF THE BRISTOL HARBOUR RAILWAY

The west portal of Redcliffe tunnel in 1981. The width accommodated double broad gauge track.  *Author*

The line emerged from the tunnel to cross Bathurst Basin. The Midland Railway, not having rail access to Bristol Harbour, had the next best thing, a non-rail connected depot opened in December 1897 at King's Wharf near Bathurst Basin, with barges linking the depot with its Avonside Wharf situated opposite Temple Meads.

The Bristol Harbour line passed behind warehouses at Bathurst Wharf and Prince's Wharf, before reaching a grain shed at Wapping Wharf near where on 30th June, 1887 four marshalling sidings were provided ending at 119 miles 45 chains. Wapping Yard had a capacity of 844 wagons. Wapping Wharf's largest crane was capable of lifting 35 tons. Private sidings continued to serve various docks, C. Hill & Sons private siding agreement being terminated 1st January, 1976.

Wapping Wharf dealt with a wide variety of traffic including coal, esparto grass, timber, sherry and Guinness. The esparto grass was the raw material for bank note paper manufacture at Wookey on the Cheddar Valley line, such trains being known as the 'Wapping and Wookey Specials'.

The single line from Redcliffe Sidings to Bathurst Bridge was worked in accordance with the electric token regulations, the token controlled by the signalman at Temple Meads Goods Yard signal box one end and by the signalman at Prince Street Crossing (119 miles 5 chains) at the other.

Bathurst Basin Bridge lifted and closed to rail. *Author's collection*

'8750' class 0-6-0PT No. 3650 (82B St Philip's Marsh) on Bathurst Basin Bridge. The broad gauge rail bearer is arrowed on the left. The water pipe to Wapping Goods Yard had to be dismantled each time the bridge was lifted. *Port of Bristol Authority.*

# DESCRIPTION OF THE BRISTOL HARBOUR RAILWAY 157

'57XX' class 0-6-0PT No. 8746 with cattle wagons at Wapping Wharf 20th August, 1960.
*R. E. Toop*

The Western Fuel Co.'s shunter at Wapping Wharf 18th August, 1980, view towards the SS *Great Britain*.
*Author*

Safety was important: before moving wagons on the dock lines and preparatory to giving enginemen a signal to move, shunters were required to walk the whole length of the train and personally caution each individual working on or near the line and a locomotive was to whistle before moving. When moving, the under-shunter had to walk in advance of the engine at such a distance that it could stop short of an obstruction, or if propelling, the head shunter was required to walk ahead of the leading vehicle. The single line between Wapping Wharf and Ashton Bridge signal box (former 'Ashton Swing Bridge North')was worked in accordance with the electric token regulations.

The Wapping Coal Distribution Centre opened on 3rd January, 1966 and continued in use until 1st May, 1987. Its own locomotive worked trains to and from Ashton Junction. *Circa* 1975 after going away for repair, it was returned by road the wrong way round. As a special dispensation it was allowed to travel by rail to the Bath Road turntable to be turned. Following closure of the line from Temple Meads in 1964, access was only via Ashton Junction via the line opened on, 4th October, 1906 alongside the New Cut. Opened as a double track line, it was reduced to single in June 1907. At Cumberland Siding ground frame, until taken out of use on 5th September, 1965, a siding served tobacco

View from Wapping Wharf towards Ashton Sidings, 18th August, 1980. The coal hoppers are from Western Fuel. *Author*

Private and not for Publication.

# GREAT WESTERN RAILWAY.

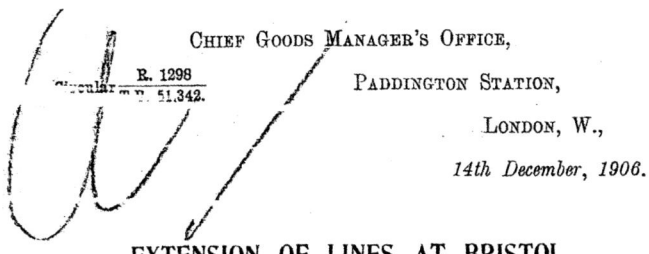

CHIEF GOODS MANAGER'S OFFICE,
PADDINGTON STATION,
LONDON, W.,
14th December, 1906.

R. 1298
Circular no. 51,342.

## EXTENSION OF LINES AT BRISTOL.

With reference to my Circular of the 2nd October, 1906. On **December 17th** further new lines, connecting with the Corporation lines, near Cumberland Basin, which give access to the Corporation Tobacco Warehouses, and the Corporation Cattle Pens, will be open for Traffic.

The rates in operation with Bristol, G.W.R. are to be applied to Traffic to and from the places referred to herein, and invoices for same, made out to Bristol, must be treated as "local" and abstracted accordingly.

The receipt of this Circular must be promptly acknowledged to your District Goods Manager.

T. H. RENDELL.

Notice of opening of lines near the Cumberland Basin.

The RCTS special near Cumberland Road Siding, 28th April, 1957.  *Dr A. J. G. Dickens*

The RCTS special approaching Ashton Swing Bridge North signal box 28th April, 1957.
*Dr A. J. G. Dickens*

# DESCRIPTION OF THE BRISTOL HARBOUR RAILWAY 161

Ashton Junction 1982. *John Mann*

'8750' class 0-6-0PT No. 3614 near Ashton Gate. *Author's collection*

Ashton Swing Bridge North signal box. The line to Canon's Marsh is on the left passing under the bridge, while that to Wapping Wharf curves sharply left behind Tobacco Bond Warehouse 'A'. The signal box door faces Bristol Corporation's tobacco warehouse siding which crosses both tracks almost at right-angles. *Michael Johnston*

A train from Wapping Wharf hauled by a '57XX' class 0-6-0PT, approaches Ashton Swing Bridge North signal box. The tobacco warehouse siding is in the foreground.
*Author's collection*

bonded warehouses. At Ashton Swing Bridge North signal box (0 miles 63 chains from Ashton Junction), the Canon's Marsh line joins. Interestingly, the box opened on 4th October, 1906 was placed in the wrong position and had to be replaced 4th November, 1906. It was renamed Ashton Bridge signal box on 1st July, 1958.

When the extension from Wapping Wharf to Ashton Junction was planned, as Bristol Corporation needed a road bridge across The Cut, it joined forces with the railway and it was decided that the bridge would have two decks: the upper carrying the road and the lower a double track railway.

Built by John Lysaght & Co. Ltd of Bristol for £70,389 its total length of 582 feet contained some 1,500 tons of steelwork, the 202 foot-long swing span alone weighing 1,000 tons. The bridge was opened by the Mayoress Mrs A. J. Smith on 3rd October, 1906 arriving by road, not rail. The swing span, which could turn either upstream or downstream, was moved by hydraulic power generated at the Underfall Yard. The two three-throw, reversible engines working the bridge were situated in a control cabin above the roadway. Two vertical shafts swung the span, only one engine being required, the other acting as a stand-by. To prevent mishaps, the bridge was interlocked with the GWR signal boxes on either side making it impossible for signals to be lowered for the passage of a train unless the swing span was firmly secured.

The double deck, road above, rail below, Ashton Swing Bridge. Partly open; this view was taken about 1920. *Author's collection*

As only one vessel was permitted to pass the bridge at a time, cones were provided to control shipping movements. For vessels moving downstream, the north cone, point upwards, was lowered, with the south cone for those wishing to proceed up river. The yard arm on which the cones were hoisted was fixed so that both cones showed when the bridge was open. The bridge master had telephone communication with two stations, one upstream and one downstream about a quarter of a mile from his bridge, warning him at night, or in foggy weather, of the approach of vessels.

When the bridge master wished to open the bridge to water traffic he rang a bell in each of the signal boxes and as soon as the signalmen had withdrawn their locking bolts this mechanically released a lever which allowed him to withdraw the bridge bolts. This movement locked the railway bolts which prevented a signalman from lowering a signal to allow a train to pass; additionally it released the lifting lever which he then operated. Immediately the ends of the bridge were correctly raised, the sliding blocks were withdrawn, the ends of the bridge lowered and the bridge swung open. When in the fully-open position a screen could be raised to obscure a red light for one direction only.

To close the bridge the screen had to be raised to reveal the red light, the bridge turned so that it was open to the railway, shooting the automatic locking bolts electrically freed the lifting lever, the bridge was raised, the sliding blocks inserted before lowering the bridge and shooting the bridge bolts thus allowing the signalmen to carry out normal working. Shipping took precedence over rail traffic and should a vessel lose a tide due to the railway blocking its passage, the railway was required to pay demurrage.

On 1st August, 1951 the Bristol Corporation Act rescinded the obligation to open it, so the bridge which had not been swung since 3rd February, 1934 was made a permanent fixture in 1953. The control cabin and road deck were removed in November 1966 when a new road system to cope with modern traffic was constructed in the area. Following the removal of the road deck, additional bracing was fitted. Under the 1897 Act, Bristol Corporation was responsible for the whole structure, the railway looking after the permanent way and paying half the cost of the bridge maintenance other than road and footpath costs. Due to corrosion, it is believed that trains have not crossed the bridge after the Festival of the Sea in May 1996.

Ashton Swing Bridge South signal box (0 miles 73 chains) also opened on 4th November, 1906 and closed on 1st July, 1958, Ashton Meadows ground frame being brought into use on that date to control the four sidings holding a total of 97 wagons. In 1961 four civil engineer's sidings

# DESCRIPTION OF THE BRISTOL HARBOUR RAILWAY

The swing bridge *circa* 1906. *Author's collection*

Ashton Swing Bridge in 1980 following the removal of the road deck. *Author*

were open close by at Bower Ashton where material such as bridge girders were stored. The sidings were closed in 1996.

Bristol Harbour Railway now runs from outside the M Shed on Prince's Wharf to the SS *Great Britain* with a branch alongside the New Cut to terminate near Ashton Swing Bridge. Although normally the preserve of steam locomotives, in August 1998 Britain's first commercial flywheel powered tram ran an experimental service. Rather than using overhead electric cables, the ultralight vehicle drew power from low-voltage electric charging points at passenger stops. This electricity powered a motor to spin a 500kg flywheel mounted horizontally under the floor. The flywheel was then able to release its energy to waft the tram along the track to the next charging-point.

The Parry railbus weighed only 6 tonnes, yet could accommodate 20 seated and 14 standing passengers. It had a range of up to two miles before it needed to recharge its 1.2 metre flywheel, though 300-500 metres was a better distance to travel and avoid a long charging time. A flywheel accepted energy much faster than batteries or any other energy storing devices during the time passengers are getting on the vehicle.

During the Second World War the line was affected by bombing: on the night of 3rd-4th January, 1941 three sheds and a granary on Princes Wharf with 8,000 tons of grain were destroyed by fire. This seven-storey building dated from 1888, was replaced by L and M sheds, later to become the Bristol Industrial Museum. High explosive bombs fell near Redcliffe tunnel and single line working was only restored on 6th January, 1941. On the night of 26th-27th February, 1941 a barrage balloon cable which had come adrift, wrapped itself round a locomotive at Wapping Wharf. On 16th-17th March, 1941 a high explosive bomb fell on No. 4 road at Wapping Wharf, also damaging Nos. 1-3 roads. On 11th-12th April, 1941 craters were formed at Wapping Wharf yard with a number of wagons derailed. On 7th-8th May, 1942 high explosive bombs blocked the Harbour line.

When a crew returned to their engine after having taken shelter during a raid, they sometimes discovered incendiary bombs lying on the flat tops of the pannier tanks.

On 1st January, 1951 the line was maintained by a permanent way gang of eight looking after the railway track plus two road men.

In 1960 a new civil engineering depot was opened parallel with Ashton Meadows sidings to replace depots at Pylle Hill and St Philip's Marsh and the bridge depots at Bath Spa and Bathampton. A brick building accommodated a workshop and stores while smaller buildings contained offices and a messroom.

Bristol Harbour Railway: SS *Great Britain* station, view west, 26th May, 2015.   *Author*

A reproduction GWR-type entrance to the SS *Great Britain* station, 26th May, 2015.   *Author*

Ashton Meadows Civil Engineer's Yard 24th February, 1961.　　　　　　　　*BR*

Ashton Meadows stores and workshops, 12th May, 1961.　　　　　　　　*BR*

# Chapter Seventeen

# The Canon's Marsh Branch

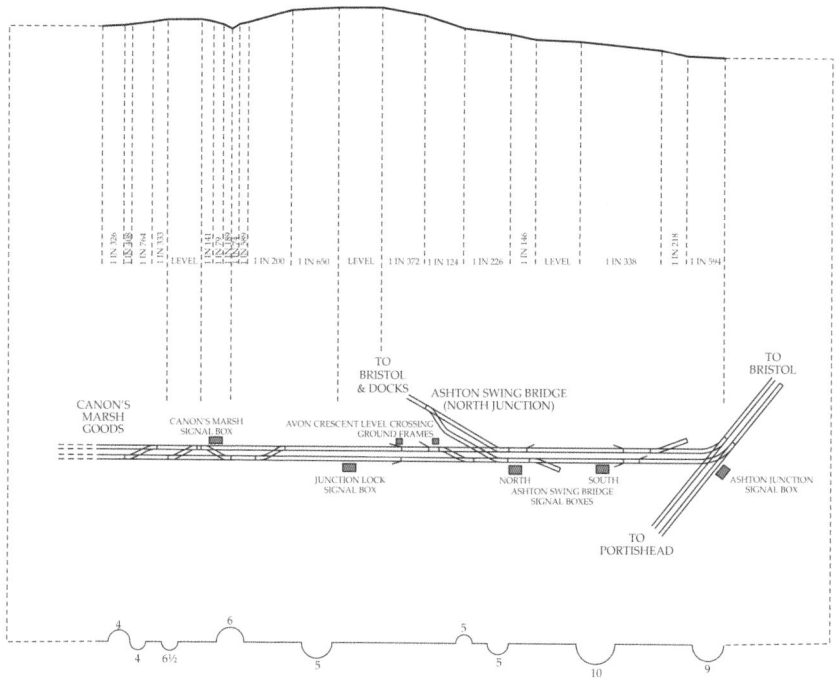

Plan of Canon's Marsh branch showing curvature and gradients.

The Bristol Harbour Railway only served the south side of the Floating Harbour – so-called because lock gates prevented water escaping thus enabling ships to float at any time and thus be independent of tides. Canon's Marsh, on the north side of the Floating Harbour, was Cathedral property and originally consisted mainly of marshy meadows stretching between the cathedral and the tide-washed banks of the Avon. Although close to the city docks, Canon's Marsh had only been slightly developed due to lack of rail communication and thus goods had to be carted or barged. The only industries were a gasworks, rope walk, lime kiln, saw mill, dry dock and a few timber yards. In order to reduce extensive transhipment, following a proposed joint GWR and Midland Railway scheme abandoned in March 1891, the GWR obtained powers under the Bristol Harbour Lines Act of 1897 to connect the Harbour Railway with the Portishead branch and lay an extension to Canon's Marsh. The main object of the extension was to give access to

The line to Wapping Wharf, right; the former line to Canon's Marsh now lifted, left. On 1st August, 1981 the photographer was standing on the site of Ashton Bridge signal box.
*John Mann*

Avon Crescent signal box. The very sharp curve has check-rails. View towards Ashton Sidings. *Michael Johnston*

THE CANON'S MARSH BRANCH 171

Avon Crescent level crossing, view towards Canon's Marsh. *Michael Johnston*

North British Type '2' B-B diesel-hydraulic D6351 near Merchants Dock, 20th November, 1963. It is standing next to Junction Lock south ground frame. *E. T. Gill*

the deep water wharves on the north side of Bristol Harbour. The GWR also planned to use the associated Bristol West Depot near Parson Street to deal with all West of England goods and so reduce congestion at Temple Meads and Pylle Hill. West Depot had accommodation for approximately 1,000 wagons plus an additional 123 at Ashton Meadows set between the West Depot and the Avon.

Numerous old houses and narrow streets were swept away and gardens in Hotwell Road had to be sacrificed to make room for the new line. Some houses in Avon Crescent were also demolished. Bristol Corporation extended the quay and built transit sheds and laid rails on its land ready for the arrival of the new line.

The line opened from Ashton Junction to Canon's Marsh 4th October, 1906, on the same date as the extension from Wapping Wharf to Ashton Junction. The opening of the line to Canon's Marsh led to development in the district. The gasworks had direct rail communication, a chocolate factory was built and a lead pipe and sheet factory rebuilt and extended, while a marble and slate importer set up an establishment.

Lairage, cattle pens and sidings were provided, chiefly for Irish traffic, Cork and Waterford steamers discharging within a few yards of the railway tracks and obviating a drive of a mile or two through the city to the railway. Rail access was also provided to a foreign animals' wharf at Merchants' Dock. In an undertaking given by the GWR in return for using land belonging to the nearby cathedral, no locomotive movements or whistling were carried out during service hours on Sundays.

In 1914 115 men were employed at the Canon's Marsh depot including 1 supervising inspector, 6 supervising foremen, 8 shunters, 20 checkers, 5 callers-off, 7 stowers, 29 porters, 36 carmen and 3 lads. The 22 sidings in the yard accommodated 531 wagons, about 100 of which were under cover, and approximately 500 wagons were exchanged daily – about the same number as the capacity of the sidings. One locomotive shunted for 22 hours daily and another for eight hours. During the Second World War LMS class '2F' 0-6-0s were on loan to the GWR and Nos. 3062 and 3064 shunted at Canon's Marsh. Transfer trips were worked from Canon's Marsh to various outlying marshalling yards set up on the periphery of Bristol.

At Bristol the GWR had two engine sheds: Bath Road for passenger locomotives and St Philip's Marsh for goods. This meant that a relief crew for a shunting engine at Canon's Marsh came from St Philip's Marsh shed and after booking on, were required to walk from the shed, following an approved route determined by locomotive inspector stepping out at 3 mph, thus crews were allowed 44 minutes for the journey of just over two miles.

A 'Dean Goods' 0-6-0 passing along Hotwell Road towards Canon's Marsh. Target No. 13 is displayed on the lower lamp bracket.
*Author's collection*

A 'Dean Goods' 0-6-0 passing along Hotwell Road.
*Author's collection*

8750' class 0-6-0PT No. 3643 proceeds along Hotwell Road with a train to Canon's Marsh, 13th January, 1965.
*Paul Strong*

Bristol Cathedral is in the background, with Canon's Marsh Goods Depot in the foreground; a 1920s view. *M. J. Tozer collection*

One of the few remaining steam-powered ships at Canon's Marsh in the mid-1950s.
*Author's collection*

The goods shed and warehouse at Canon's Marsh measured 540 feet by 133 feet. It was of ferro-concrete on the Hennebique principle, there being 274 ferro-concrete piles averaging 32 feet in depth able to accept floor loadings up to six hundredweights per square foot. The shed was 35 feet high and had a floor area of 35,600 square feet. The steel and corrugated-iron roof had Rendle's patent glazing. The contractor responsible for building was Samuel Robertson of Bristol, the work being carried out under the supervision of P. E. Culverhouse.

Four roads holding about 97 wagons passed through the building, with a platform 20-feet wide between the two most northern lines and one of the same dimensions between the southern line and cartway. The depot was equipped with eight electric cranes and three 30 hundredweight electric hoists. The largest crane had a lift of 30 tons. The upper storey of the shed was a warehouse and had a wood-block floor. Electric hoists raised traffic from the platforms below. In 1925 two of the existing electric platform cranes were given an extension to gain a greater height of lift and a 30-ton electric gantry crane was provided in the yard to replace a 12-ton fixed crane.

Shunting on the quay required the utmost vigilance as lorries might obstruct the lines, dockers were going about their duties and cranes were swinging loads. Before shunting on the quay, a driver had to remember to raise and clip the Automatic Train Control apparatus to prevent it being fouled.

On 15th May, 1952 a van belonging to William Butler, tar distillers, collided with a locomotive at Merchants' Dock level crossing. The van, caught by one of the engine's footsteps, was dragged 20 yards along the track and exploded when crushed against a wall, killing the driver. Immediately the driver saw the van he braked, but with 43 wagons behind him, could not stop immediately.

Following this fatality the suggestion was put forward that locomotives be fitted with warning bells. BR said that it would be necessary to fit 12 engines with bells for the six booked services and to cover every locomotive likely to be used 32 would have to be fitted. As drivers had been instructed to keep a good look-out, it was believed that bells were unnecessary.

Although most railwaymen were honest, a few fell to temptation. During the Second World War one shunter deliberately let a van containing sugar, then rationed, go hard up against a stop block knowing that something would break, thus allowing him to place a 'Not to Go' label on the van so that he could shunt it to the cripples' siding where he and his mates could help themselves to its contents.

Due to the rundown of the city docks, the Canon's Marsh branch closed on 14th June, 1965. Bristol Corporation's Parliamentary Bill for closing the city docks passed its final reading on 15th July, 1971.

Engines in regular use on the Canon's Marsh pilot in 1943-1944 were: Nos. 2015, 2031, 2064, 2070, 2135. The '16XX', a modern version of the '2021' class, first appeared at Canon's Marsh in May 1951 when allocated ex-works to St Philip's Marsh. In the 1960s came the D20XX 0-6-0 204 bhp shunters, later classified '03'.

One unusual engine working the line was D0226, an English Electric 0-6-0 500 bhp shunter. It remained the property of the manufacturer, but BR provided facilities for road tests.

*Above*: '2021' class 0-6-0PT No. 2070 (82B St Philip's Marsh) at Canon's Marsh, July 1952. Behind the engine is a shunters' truck. Because the locomotive travels backwards and forwards frequently, it displays both a white light and a red light.
*M. J. Tozer*

*Right*: No. 3643 approaches Canon's Marsh 13th January, 1965. Notice the fireman increasing the tractive effort!
*Paul Strong*

# Chapter Eighteen

## Description of the Canon's Marsh Branch

From a junction with the Bristol Harbour Railway at Ashton Swing Bridge North signal box (0 miles 63 chains) the double track line reached Avon Crescent signal box (0 miles 8 chains from the junction) which protected a level crossing. This signal box obtained its water from Ashton Swing Bridge box – this was rare as a box usually obtained water from a station rather than the next box. The railway crossed Old Junction Lock by a fixed bridge before reaching Junction Lock ground frame (0 miles 12 chains) protecting one end of the hydraulically-operated swing bridge over Junction Lock. Initially protected at its northern end by Junction Lock signal box (0 miles 16 chains), as an economy measure the box was converted to Junction Lock North ground frame on 18th March, 1925. Here the line reached Merchants' Dock, filled in in 1966.

Then followed several level crossings: Denty's Timber Yard, Merchants' Dock, Blackhorse Lane, Dock Gates and Poole's Siding and beyond the latter the line was sandwiched between Hotwell Road and the Floating Harbour. It was not uncommon when reaching HMS *Flying Fox* for the fireman have to go aboard to seek out those of its crew who had obstructed the line with their cars. When re-starting a driver would be wise to check that his brake van was still attached to the train as some Hotwells youngsters sought entertainment by uncoupling wagons. The

'57XX' class 0-6-0PT No. 9769 heading an RCTS special on 26th September, 1959, passes the training ship *Flying Fox* on its way to Canon's Marsh.  *Dr A. J. G. Dickens*

177

line was supported along the Floating Harbour by a concrete retaining wall 500-feet long, with an a average width of 10 feet.

One morning *circa* 1937 as No. 5 transfer consisting of a '2301' class 0-6-0 and brake van were travelling along the quayside the crew was enjoying breakfast fried on a shovel. About halfway along Hotwells Road a lorry belonging to D. M. W. Bullock Limited crossed the rails and started to unload beer on to the P. & A. Campbell's paddle steamer *Glen Usk*. Before the engine driver could take action, he had pushed the lorry and contents into the Floating Harbour.

Canon's Marsh signal box (0 miles 61 chains) was near the entrance to the gas works' private sidings which formed an oval. Canon's Marsh box had a minute frame with three small electric levers only about a foot high. The box too was unusually small, measuring about 12 feet in length and 4 feet wide, in fact inside men could only pass by turning sideways. The two banner disc signals – one for entering and the other for leaving – were very unusual in that they were lit by gas, but moved electrically.

Traffic for the gas works was required to be worked to Canon's Marsh Yard and then worked back by the shunting engine. Traffic within the gas works was moved by the company's own engine. Straight ahead was Canon's Marsh Goods Depot a mile from Ashton Swing Bridge, while lines to serve quays on the Floating Harbour curved round behind the gas works. Trains in the Canon's Marsh dock area were restricted to a maximum speed of 4 mph. The maximum load between Ashton Junction and Canon's Marsh was 32 wagons of ordinary length and a brake van, though between 4.00 pm and 6.00 am any train not exceeding 45 wagons and a brake van was allowed to leave Canon's Marsh if it had been ascertained that the Junction Lock Swing Bridge was in the rail position.

During the Second World War the line was hit several times. On 2nd December, 1940 the shed at Canon's Marsh goods depot and several wagons were set on fire by incendiary bombs and the branch closed by an unexploded bomb at Avon Crescent. On 3rd-4th January, 1941 the line was blocked by direct hit near Avon Crescent that damaged the bridge over the lock. On 11th-12th April, 1941 serious damage was done to the up and down lines to Canon's Marsh and to the Ashton Meadows exchange sidings which were not cleared until 5.00 pm on 17th April; craters were also formed in Canon's Marsh goods yard. On 7th-8th May 1942 the Canon's Marsh line was blocked by debris.

On 1st January, 1951 the Canon's Marsh and Wapping Wharf lines were maintained by a permanent way gang of seven plus two road men.

# DESCRIPTION OF THE CANON'S MARSH BRANCH 179

No. 9769 runs round the RCTS special at Canon's Marsh 26th September, 1959. The gas works buildings are on the right. *Dr A. J. G. Dickens*

A view of Canon's Marsh 1st August, 1981. *John Mann*

## Appendix One

### TRAFFIC DEALT WITH

| STATION. | YEAR. | STAFF. | | TOTAL RECEIPTS. | PASSENGER TRAIN TRAFFIC. | | | | | |
|---|---|---|---|---|---|---|---|---|---|---|
| | | Supervisory and Wages (all Grades). | Paybill Expenses. | | Tickets issued. | Season Tickets. | Passengers (including Season Tickets, etc.) | Receipts. | | |
| | | | | | | | | Parcels. | Miscellaneous | Total. |
| | | No. | £ | £ | No. | No. | £ | £ | £ | £ |
| Portishead Branch. | | | | | | | | | | |
| Canon's Marsh | | | | | No Passenger Traffic dealt with. | | | | | |
| Ashton Gate Platform. | | | | | Included with Clifton Bridge prior to 1926 | | | | | |
| | 1926 | Included | | 396 | 9,013 | — | 396 | — | — | 396 |
| | 1927 | with | | 697 | 20,402 | — | 697 | — | — | 697 |
| | 1928 | Clifton Bridge. | | 1,106 | 28,579 | — | 1,106 | — | — | 1,106 |
| | 1929 | | | 1,548 | 39,616 | — | 1,527 | 17 | 4 | 1,548 |
| | 1930 | | | 1,460 | 36,384 | 7 | 1,420 | 35 | 5 | 1,460 |
| | 1931 | | | 1,361 | 33,019 | 14 | 1,308 | 48 | 5 | 1,361 |
| | 1932 | | | 1,234 | 30,560 | 12 | 1,194 | 37 | 3 | 1,234 |
| | 1933 | | | 1,299 | 33,077 | 13 | 1,246 | 50 | 3 | 1,299 |
| Clifton Bridge | 1903 | 10 | 443 | 5,343 | 85,336 | * | 2,218 | 33 | 16 | 2,267 |
| | 1913 | 6 | 411 | 4,865 | 101,403 | * | 2,616 | 54 | 123 | 2,793 |
| | 1923 | 7 | 1,008 | 9,043 | 89,252 | 82 | 3,606 | 126 | — | 3,732 |
| | 1924 | 6 | 1,004 | 7,305 | 68,365 | 46 | 2,676 | 110 | 13 | 2,799 |
| | 1925 | 6 | 1,060 | 6,379 | 74,011 | 27 | 2,730 | 73 | 45 | 2,848 |
| | 1926 | 6 | 1,068 | 5,101 | 59,014 | 27 | 2,152 | 62 | 18 | 2,232 |
| | 1927 | 7 | 1,192 | 7,135 | 82,526 | 21 | 2,677 | 51 | 41 | 2,769 |
| | 1928 | 6 | 1,227 | 7,670 | 101,735 | 9 | 3,248 | 73 | 34 | 3,355 |
| | 1929 | 6 | 1,275 | 7,073 | 78,016 | 21 | 2,560 | 74 | 34 | 2,668 |
| | 1930 | 8 | 1,333 | 7,568 | 60,358 | 25 | 2,034 | 68 | 59 | 2,161 |
| | 1931 | 8 | 1,324 | 6,958 | 43,474 | 25 | 1,537 | 64 | 444 | 2,045 |
| | 1932 | 8 | 1,258 | 5,392 | 34,882 | 21 | 1,273 | 53 | 56 | 1,382 |
| | 1933 | 8 | 1,289 | 5,757 | 38,192 | 9 | 1,269 | 58 | 39 | 1,366 |
| Ham Green Halt. | | | | Opened | December, 1926. | | | | | |
| | 1927 | Included | | Traffic included with Pill. | | | | | | |
| | 1928 | with | | 2 | 59 | — | 2 | — | — | 2 |
| | 1929 | Pill. | | 4 | 139 | — | 4 | — | — | 4 |
| | 1930 | | | 4 | 114 | — | 4 | — | — | 4 |
| | 1931 | | | 2 | 92 | — | 2 | — | — | 2 |
| | 1932 | | | 3 | 60 | — | 3 | — | — | 3 |
| | 1933 | | | 2 | 56 | — | 2 | — | — | 2 |
| Nightingale Valley Halt. | | | | Opened July, 1928. Included with Clifton Bridge. | | | | | | |
| | | | | Closed September, 1932. | | | | | | |
| Pill .. .. .. | 1903 | 4 | 251 | 2,959 | 68,021 | * | 1,732 | 75 | 11 | 1,818 |
| | 1913 | 5 | 295 | 2,963 | 78,436 | * | 2,060 | 66 | 3 | 2,129 |
| | 1923 | 6 | 1,005 | 4,789 | 83,379 | 535 | 3,712 | 112 | 34 | 3,858 |
| | 1924 | 6 | 987 | 3,977 | 62,537 | 372 | 2,836 | 104 | 33 | 2,973 |
| | 1925 | 6 | 993 | 3,944 | 63,506 | 346 | 2,890 | 96 | 71 | 3,057 |
| | 1926 | 6 | 938 | 3,421 | 51,336 | 309 | 2,348 | 101 | 38 | 2,487 |
| | 1927 | 6 | 984 | 3,259 | 58,760 | 280 | 2,411 | 107 | 18 | 2,536 |
| | 1928 | 6 | 955 | 3,370 | 70,980 | 260 | 2,616 | 95 | 7 | 2,718 |
| | 1928 | 8 | 1,107 | 3,220 | 65,976 | 240 | 2,465 | 105 | 26 | 2,596 |
| | 1930 | 8 | 1,279 | 3,303 | 65,285 | 234 | 2,408 | 108 | 12 | 2,528 |
| | 1931 | 8 | 1,261 | 2,897 | 58,833 | 262 | 2,156 | 105 | 2 | 2,263 |
| | 1932 | 8 | 1,261 | 2,540 | 51,488 | 162 | 1,909 | 85 | 5 | 1,999 |
| | 1933 | 7 | 1,110 | 2,561 | 48,965 | 160 | 1,704 | 79 | 1 | 1,874 |

* Not available.

Traffic Dealt with at Branch Stations 1903 – 1933

# TRAFFIC DEALT WITH AT BRANCH STATIONS

**BRISTOL DIVISION.**
(Branch Lines.)

| | | | | | | | | | | |
|---|---|---|---|---|---|---|---|---|---|---|
| | \multicolumn{3}{c}{Forwarded.} | | \multicolumn{3}{c}{Received.} | | | | |
| | \multicolumn{8}{c}{GOODS TRAIN TRAFFIC.} | | | |

| Coal and Coke "Charged." | Other Minerals. | General Merchandise. | Coal and Coke "Charged." | Other Minerals. | General Merchandise. | Coal and Coke "Not Charged" (Forwarded and Received). | Total Goods Tonnage. | Total Receipts (excluding "Not Charged" Coal and Coke). | Livestock (Forwarded and Received). | Total Carted Tonnage (included in Total Goods Tonnage). |
|---|---|---|---|---|---|---|---|---|---|---|
| Tons. | Tons. | Tons. | Tons. | Tons. | Tons. | Tons. | Tons. | £ | Wagons. | Tons. |
| | | | \multicolumn{6}{l}{Included with Bristol (Temple Meads).} | | | |
| | | | \multicolumn{6}{l}{No Goods Traffic dealt with.} | | | |
| 4,546 | 5,260 | 1,838 | 725 | 2,982 | 635 | 370 | 16,356 | 3,076 | — | 78 |
| 94 | 6,491 | 383 | 1,209 | 3,584 | 180 | 2,771 | 14,712 | 2,072 | 23 | 91 |
| 10 | 674 | 718 | 1,109 | 1,381 | 6,675 | 814 | 11,381 | 5,311 | — | 156 |
| — | 160 | 528 | 1,257 | 953 | 6,446 | 578 | 9,922 | 4,506 | — | 140 |
| 18 | 217 | 696 | 1,247 | 963 | 7,191 | 1,194 | 11,526 | 3,531 | — | 146 |
| — | 119 | 644 | 919 | 762 | 4,141 | 477 | 7,062 | 2,869 | — | 105 |
| — | 358 | 1,124 | 1,629 | 864 | 8,057 | 679 | 13,811 | 4,366 | — | 173 |
| — | 244 | 1,375 | 1,278 | 1,061 | 8,735 | 757 | 13,450 | 4,315 | — | 146 |
| — | 280 | 1,570 | 1,310 | 690 | 7,512 | 1,016 | 12,378 | 4,405 | 3 | 155 |
| 15 | 413 | 1,867 | 1,497 | 277 | 12,102 | 883 | 17,054 | 5,407 | 4 | 84 |
| 8 | 85 | 1,346 | 1,781 | 23 | 9,976 | 330 | 13,549 | 4,913 | 5 | 85 |
| — | 67 | 853 | 1,220 | 170 | 8,862 | 128 | 11,300 | 4,010 | 1 | 79 |
| — | 329 | 1,305 | 1,085 | 36 | 9,586 | 90 | 13,031 | 4,391 | 2 | 383 |
| — | 32 | 847 | 1,815 | 2,531 | 1,117 | 1,950 | 8,292 | 1,141 | — | 179 |
| — | 9 | 528 | 1,798 | 850 | 819 | 2,130 | 6,134 | 834 | — | 206 |
| 7 | 47 | 165 | 1,818 | 99 | 240 | 3,331 | 5,707 | 931 | — | 121 |
| 40 | 89 | 145 | 1,963 | 142 | 254 | 2,009 | 4,642 | 1,004 | 8 | 123 |
| — | 45 | 62 | 1,337 | 464 | 346 | 2,353 | 4,607 | 887 | 10 | 122 |
| — | 26 | 104 | 1,020 | 234 | 398 | 2,160 | 3,942 | 934 | 9 | 156 |
| 9 | 61 | 70 | 1,166 | 17 | 308 | 3,514 | 5,175 | 723 | 8 | 128 |
| — | — | 55 | 1,119 | 31 | 255 | 2,793 | 4,253 | 652 | 10 | 113 |
| — | — | 62 | 1,411 | 24 | 156 | 2,750 | 4,403 | 624 | 17 | 69 |
| — | 10 | 8 | 27 | 1,066 | 666 | 302 | 2,965 | 5,044 | 775 | 28 | 74 |
| — | — | — | 22 | 967 | 44 | 160 | 3,274 | 4,476 | 434 | 11 | 69 |
| — | — | 6 | 25 | 1,013 | 15 | 175 | 3,185 | 4,419 | 541 | 3 | 66 |
| — | — | — | 52 | 825 | 154 | 292 | 2,794 | 4,117 | 687 | 8 | 90 |

## TRAFFIC DEALT WITH

| STATION. | YEAR. | STAFF. | | TOTAL RECEIPTS. | PASSENGER TRAIN TRAFFIC. | | | | | |
|---|---|---|---|---|---|---|---|---|---|---|
| | | Supervisory and Wages (all Grades). | Paybill Expenses. | | Tickets issued. | Season Tickets. | Receipts. | | | |
| | | | | | | | Passengers (including Season Tickets, etc.) | Parcels. | Miscellaneous | Total. |
| | | No. | £ | £ | No. | No. | £ | £ | £ | £ |
| Portbury (†) | 1903 | 2 | 125 | 1,189 | 20,069 | * | 593 | 35 | 247 | 875 |
| | 1913 | 2 | 130 | 1,424 | 17,724 | * | 604 | 59 | 463 | 1,126 |
| | 1923 | 5 | 714 | 2,777 | 16,946 | 124 | 883 | 27 | 571 | 1,481 |
| | 1924 | 4 | 675 | 2,382 | 12,224 | 105 | 673 | 34 | 608 | 1,315 |
| | 1925 | 4 | 674 | 2,516 | 10,600 | 96 | 650 | 26 | 929 | 1,605 |
| | 1926 | 4 | 621 | 2,715 | 8,665 | 60 | 527 | 23 | 883 | 1,433 |
| | 1927 | 5 | 756 | 1,707 | 13,254 | 44 | 565 | 28 | 703 | 1,296 |
| | 1928 | 5 | 807 | 1,640 | 14,765 | 25 | 582 | 27 | 533 | 1,142 |
| | 1929 | 5 | 810 | 1,450 | 12,484 | 36 | 519 | 26 | 398 | 943 |
| | 1930 | 5 | 819 | 1,109 | 11,855 | 45 | 459 | 31 | 246 | 736 |
| | 1931 | 4 | 590 | 858 | 10,187 | 40 | 367 | 43 | 109 | 519 |
| | 1932 | 4 | 566 | 742 | 9,713 | 27 | 342 | 47 | 88 | 477 |
| | 1933 | 4 | 569 | 656 | 7,814 | 26 | 302 | 47 | 93 | 442 |
| Portbury Ship Yard Halt. | | | | Included | with Portbury. | | | | | |
| Portishead | 1903 | 7 | 400 | 7,022 | 74,207 | * | 4,068 | 478 | 97 | 4,643 |
| | 1913 | 10 | 752 | 6,818 | 82,077 | * | 4,221 | 301 | 132 | 4,654 |
| | 1923 | 25 | 3,922 | 13,867 | 83,975 | 949 | 8,898 | 742 | 252 | 9,892 |
| | 1924 | 25 | 4,383 | 12,119 | 79,021 | 917 | 7,896 | 372 | 278 | 8,546 |
| | 1925 | 25 | 4,511 | 12,201 | 85,113 | 839 | 7,910 | 359 | 320 | 8,589 |
| | 1926 | 24 | 3,825 | 10,971 | 74,280 | 839 | 6,686 | 371 | 276 | 7,333 |
| | 1927 | 29 | 4,260 | 22,088 | 82,657 | 819 | 7,238 | 443 | 312 | 7,993 |
| | 1928 | 30 | 4,773 | 25,665 | 80,787 | 874 | 8,126 | 546 | 191 | 8,863 |
| | 1929 | 29 | 4,535 | 19,669 | 97,418 | 1,019 | 8,005 | 559 | 205 | 8,769 |
| | 1930 | 25 | 4,240 | 17,243 | 96,711 | 1,152 | 7,745 | 482 | 151 | 8,378 |
| | 1931 | 25 | 3,899 | 14,608 | 87,596 | 1,043 | 6,733 | 422 | 178 | 7,333 |
| | 1932 | 23 | 3,404 | 14,826 | 74,615 | 916 | 6,187 | 420 | 114 | 6,721 |
| | 1933 | 20 | 3,233 | 11,710 | 73,011 | 783 | 5,640 | 418 | 38 | 6,096 |
| Portishead Docks. (†) | 1903 | | | 19,664 | | | | | | |
| | 1913 | | | 53,336 | | | | | | |
| | 1923 | | | 157,241 | No Passenger Traffic dealt with. | | | | | |
| | 1924 | Included with Portishead. | | 177,534 | | | | | | |
| | 1925 | | | 180,728 | | | | | | |
| | 1926 | | | 116,962 | | | | | | |
| | 1927 | | | 106,354 | | | | | | |
| | 1928 | | | 89,180 | | | | | | |
| | 1929 | | | 71,096 | | | | | | |
| | 1930 | | | 63,355 | | | | | | |
| | 1931 | | | 14,517 | | | | | | |
| | 1932 | | | 9,510 | | | | | | |
| | 1933 | | | 7,891 | | | | | | |
| Total | 1903 | 23 | 1,219 | 36,177 | 247,633 | * | 8,611 | 621 | 371 | 9,603 |
| | 1913 | 23 | 1,588 | 69,406 | 279,640 | * | 9,501 | 480 | 721 | 10,702 |
| | 1923 | 43 | 6,649 | 187,717 | 273,552 | 1,600 | 17,099 | 1,007 | 857 | 18,963 |
| | 1924 | 41 | 7,049 | 203,317 | 222,147 | 1,440 | 14,081 | 620 | 932 | 15,633 |
| | 1925 | 41 | 7,228 | 205,768 | 233,230 | 1,308 | 14,180 | 554 | 1,365 | 16,099 |
| | 1926 | 40 | 6,452 | 139,566 | 202,308 | 1,235 | 12,109 | 557 | 1,215 | 13,881 |
| | 1927 | 47 | 7,192 | 141,240 | 257,599 | 1,164 | 13,588 | 629 | 1,074 | 15,291 |
| | 1928 | 47 | 7,762 | 128,633 | 305,905 | 1,108 | 15,680 | 741 | 765 | 17,186 |
| | 1929 | 48 | 7,727 | 104,060 | 293,649 | 1,316 | 15,080 | 781 | 667 | 16,528 |
| | 1930 | 46 | 7,671 | 94,042 | 270,707 | 1,463 | 14,070 | 724 | 473 | 15,267 |
| | 1931 | 45 | 7,074 | 41,001 | 233,201 | 1,384 | 12,103 | 682 | 738 | 13,523 |
| | 1932 | 43 | 6,489 | 34,247 | 201,327 | 1,138 | 10,908 | 642 | 266 | 11,816 |
| | 1933 | 39 | 6,201 | 29,876 | 201,115 | 1,000 | 10,253 | 652 | 174 | 11,079 |

\* Not available.   † Controlled by Portishead.

## AT STATIONS.

**BRISTOL DIVISION.**
(Branch Lines.

### GOODS TRAIN TRAFFIC.

| Forwarded. | | | Received. | | | Coal and Coke "Not Charged" (Forwarded and Received). | Total Goods Tonnage. | Total Receipts (excluding "Not Charged" Coal and Coke). | Livestock (Forwarded and Received). | Total Carted Tonnage (included in Total Goods Tonnage). |
|---|---|---|---|---|---|---|---|---|---|---|
| Coal and Coke "Charged." | Other Minerals. | General Merchandise. | Coal and Coke "Charged." | Other Minerals. | General Merchandise. | | | | | |
| Tons. | Tons. | Tons. | Tons. | Tons. | Tons. | Tons. | Tons. | £ | Wagons. | Tons. |
| — | 34 | 73 | 518 | 895 | 107 | 243 | 1,870 | 314 | 18 | 97 |
| — | — | 235 | 332 | 69 | 661 | 245 | 1,542 | 298 | 15 | 81 |
| — | 459 | 453 | 193 | 113 | 244 | 322 | 1,784 | 1,296 | 62 | 142 |
| 27 | 317 | 496 | 179 | 195 | 258 | 349 | 1,821 | 1,067 | 64 | 383 |
| — | 177 | 287 | 149 | 148 | 342 | 312 | 1,415 | 911 | 80 | 70 |
| 10 | 455 | 778 | 58 | 4 | 348 | 145 | 1,798 | 1,282 | 73 | 105 |
| 8 | — | 50 | 192 | 82 | 248 | 206 | 876 | 411 | 45 | 50 |
| — | — | 36 | 304 | 72 | 244 | 234 | 890 | 498 | 63 | 44 |
| — | — | 19 | 316 | 34 | 220 | 101 | 780 | 507 | 66 | 25 |
| 2 | — | 28 | 177 | 6 | 187 | 170 | 570 | 373 | 58 | 24 |
| — | — | 23 | 42 | 35 | 154 | 208 | 462 | 339 | 58 | 31 |
| — | 20 | 26 | 69 | 78 | 131 | 167 | 491 | 265 | 30 | 22 |
| — | — | 24 | 32 | 6 | 141 | 175 | 378 | 214 | 23 | 22 |
| | | | | | | | | | | |
| 84 | 728 | 408 | 2,395 | 2,707 | 1,921 | 2,731 | 10,974 | 2,379 | 21 | 583 |
| 310 | 14 | 602 | 1,734 | 637 | 2,722 | 3,852 | 9,871 | 2,164 | 38 | 819 |
| 779 | 321 | 660 | 1,760 | 9,906 | 1,752 | 4,152 | 19,339 | 8,975 | 20 | 1,082 |
| 733 | 157 | 580 | 2,028 | 7,675 | 1,903 | 4,378 | 17,454 | 3,573 | 22 | 1,119 |
| 795 | 188 | 528 | 1,498 | 6,260 | 1,798 | 4,411 | 15,478 | 3,612 | 18 | 1,068 |
| 600 | 222 | 585 | 2,137 | 679 | 1,855 | 2,413 | 8,491 | 3,638 | 18 | 1,045 |
| 600 | 33,942 | 820 | 2,491 | 6,472 | 3,618 | 5,807 | 58,840 | 14,095 | 16 | 1,895 |
| 841 | 5,581 | 1,065 | 3,305 | 12,107 | 12,550 | 6,269 | 41,778 | 16,802 | 23 | 1,837 |
| 909 | 267 | 853 | 2,695 | 9,838 | 7,936 | 6,397 | 28,895 | 10,900 | 25 | 1,394 |
| 639 | 306 | 1,449 | 2,016 | 25,990 | 3,582 | 4,773 | 38,746 | 8,865 | 13 | 1,667 |
| 706 | 237 | 1,006 | 2,210 | 31,907 | 2,381 | 4,911 | 44,258 | 7,275 | 13 | 2,339 |
| 987 | 213 | 1,820 | 2,179 | 48,588 | 2,425 | 4,474 | 60,086 | 8,105 | 11 | 2,069 |
| 819 | 192 | 656 | 2,452 | 44,581 | 1,703 | 4,043 | 55,046 | 5,615 | 7 | 1,294 |
| | | | | | | | | | | |
| — | 13,833 | 57,882 | 1,001 | 4,356 | 507 | 18 | 77,597 | 19,664 | — | 1,042 |
| — | — | 117,301 | 1,755 | 189 | 13,128 | 500 | 132,873 | 53,336 | — | 17,560 |
| 71 | 1,118 | 157,962 | 274 | 846 | 7,564 | 7,545 | 175,380 | 157,241 | — | 4,266 |
| 6 | 1,294 | 207,045 | 250 | 641 | 7,199 | 6,758 | 223,193 | 177,534 | — | 4,353 |
| — | 2,071 | 214,236 | 218 | 696 | 5,170 | 4,806 | 227,206 | 180,728 | — | 3,794 |
| — | 1,448 | 151,593 | 175 | 608 | 3,127 | 2,007 | 158,958 | 116,962 | — | 3,003 |
| 857 | 1,777 | 128,992 | 124 | 34,043 | 7,807 | 3,913 | 177,513 | 106,354 | — | 2,304 |
| — | 1,239 | 118,691 | 114 | 5,787 | 6,952 | 2,976 | 135,759 | 89,180 | — | 3,847 |
| 96 | 928 | 90,569 | 78 | 800 | 6,607 | 3,061 | 102,139 | 71,096 | — | 2,337 |
| — | 503 | 89,976 | 65 | 1,044 | 7,611 | 3,716 | 102,915 | 63,355 | — | 1,909 |
| — | 742 | 31,032 | 90 | 751 | 7,743 | 2,127 | 42,485 | 14,517 | — | 434 |
| — | 607 | 18,034 | 44 | 1,074 | 9,563 | 2,151 | 31,473 | 9,510 | — | 1,504 |
| 11 | 1,363 | 12,577 | 94 | 989 | 8,160 | 1,749 | 25,943 | 7,891 | — | 2,840 |
| | | | | | | | | | | |
| 4,630 | 19,887 | 61,048 | 6,454 | 13,471 | 4,287 | 5,312 | 115,089 | 26,574 | 39 | 1,979 |
| 404 | 6,514 | 119,049 | 6,828 | 5,329 | 17,510 | 9,498 | 165,132 | 58,704 | 76 | 18,757 |
| 867 | 2,619 | 159,967 | 5,154 | 12,345 | 16,475 | 16,164 | 213,591 | 168,754 | 82 | 5,767 |
| 806 | 2,017 | 208,794 | 5,677 | 9,606 | 16,060 | 14,072 | 257,032 | 187,684 | 94 | 6,118 |
| 813 | 2,698 | 215,809 | 4,449 | 8,531 | 14,856 | 13,076 | 260,232 | 189,669 | 117 | 5,200 |
| 610 | 2,270 | 153,704 | 4,309 | 2,287 | 9,869 | 7,202 | 180,251 | 125,685 | 100 | 4,504 |
| 1,564 | 36,138 | 131,056 | 5,632 | 41,478 | 25,638 | 14,209 | 255,715 | 125,949 | 69 | 4,555 |
| 841 | 7,064 | 121,222 | 6,180 | 19,058 | 28,736 | 13,029 | 196,130 | 111,447 | 96 | 5,987 |
| 1,005 | 1,475 | 93,073 | 5,810 | 11,386 | 22,431 | 13,415 | 148,595 | 87,532 | 111 | 3,080 |
| 666 | 1,220 | 93,338 | 4,821 | 27,983 | 23,784 | 12,516 | 164,338 | 78,775 | 101 | 3,758 |
| 714 | 1,004 | 34,329 | 5,090 | 32,760 | 20,423 | 10,850 | 105,230 | 27,478 | 87 | 2,058 |
| 987 | 913 | 20,758 | 4,525 | 40,925 | 21,156 | 10,105 | 108,369 | 22,431 | 54 | 3,740 |
| 830 | 1,884 | 15,614 | 5,088 | 45,766 | 19,882 | 9,451 | 98,515 | 18,797 | 40 | 4,629 |

# Appendix Two

## Industrial Locomotives

### Portishead Docks

Normally only two were stabled at Portishead, initially in a single-road shed near the entrance lock, but in 1951 this was replaced by a two-road shed. This closed *circa* 1970 and locomotives were then stabled in the open air.

The Port of Bristol issued its own rule book for the movement of trains and locomotives and no engine was allowed on British Railways tracks unless it carried a British Railways registration plate.

| Name | Wheel arrangement | Builder | Works Number | Year | Source | Disposal |
|---|---|---|---|---|---|---|
| *Alexander* | 0-6-0ST | Fox, Walker | 280 | 1875 | New | Scrapped *circa* 1934 |
| *Harold* | 0-6-0ST | Peckett | 459 | 1887 | New | Scrapped 1949 |
| *Leslie* | 0-6-0ST | Avonside | 1371 | 1898 | New | Scrapped 1933 |
| *William* | 0-6-0ST | Avonside | 1725 | 1915 | New | To Avonmouth |
| *Lionel* | 0-6-0ST | Peckett | 466 | 1889 | From Avonmouth | To Avonmouth |
| *Kenneth* | 0-6-0ST | Peckett | 808 | 1900 | From Avonmouth | To Avonmouth |
| *Alfred* | 0-6-0PT | Avonside | 1679 | 1914 | From Avonmouth | To Avonmouth |
| *Murray* | 0-6-0PT | Peckett | 1006 | 1904 | From Avonmouth | Scrapped 1958 |
| D2001 *Norman* | 0-4-0DM | Hudswell, Clarke | D744 | 1950 | From Avonmouth | To Avonmouth |
| D1001 *Gordano* | 0-4-0DM | Hudswell, Clarke | D894 | 1954 | New | Scrapped 1973 |
| 26 *Douglas* | 0-6-0DM | Hudswell, Clarke | D916 | 1956 | From Avonmouth | To Avonmouth |
| 30 | 0-6-0DM | Hudswell, Clarke | D1171 | 1959 | From Avonmouth | To Avonmouth |
| 31 | 0-6-0DM | Hudswell, Clarke | D1172 | 1959 | From Avonmouth | To Avonmouth |
| 37 | 0-6-0DM | Sentinel | 10151 | 1964 | From Avonmouth | To Avonmouth |

*Harold* and *Kenneth* were both registered to work over GWR tracks receiving the registrations numbers 191 and 192 respectively.

### Portishead Power Station

The private siding agreement was dated 24th August, 1955 and terminated 31st December, 1967.

| Name | Wheel arrangement | Builder | Works Number | Year | Source | Disposal |
|---|---|---|---|---|---|---|
| No. 7 | 0-4-0DM | John Fowler | 4210144 | 1958 | New | To Esso, Purfleet *circa* 1962 |
| No. 6 | 0-4-0DM | John Fowler | 4210143 | 1958 | ex-Purfleet Storage | To George Cohen, Cransley Depot, Kettering |

# INDUSTRIAL LOCOMOTIVES

## Albright & Wilson, Portishead

| Name | Wheel arrangement | Builder | Works Number | Year | Source | Disposal |
|---|---|---|---|---|---|---|
| | 0-4-0DM | Peckett | 5000 | 1956 | Hired from Peckett | 1957 returned |
| | 0-4-0DM | Peckett | 5002 | 1957 | New | c.1971 scrapped |
| | 0-4-0ST | Peckett | 1611 | 1923 | Hired from Peckett 1959 | To Cornish Steam Loco Society 1978 |
| D2001 *Norman* | 0-4-0DM | Hudswell, Clarke | D744 | 1950 | ex-Avonmouth Docks | Scrapped July 1973 |
| | 0-4-0DE | Ruston & Hornsby | 38175 | 1955 | ex-Thomas Ward | To Oldbury Works |

## Ashton Containers Ltd

| Name | Wheel arrangement | Builder | Works Number | Year | Source | Disposal |
|---|---|---|---|---|---|---|
| | 0-4-0DM | John Fowler | 22288 | 1938 | New | To Joseph Pugsley 1965 |

Western Fuel Co's 0-6-0 diesel-mechanical shunter at Wapping Wharf 18th August, 1980. It was originally the Port of Bristol Authority's No 30, built by Hudswell, Clarke & Co.
*Author*

## William Cowlin & Son's Portishead Power Station Contract.

Standard gauge siding connection with GWR opened 25th March, 1927; power station opened February 1929.

| Name | Wheel arrangement | Builder | Works Number | Year | Source | Disposal |
|---|---|---|---|---|---|---|
| Portishead | 0-6-0ST | Manning, Wardle | 1134 | 1890 | ex-WCPLR | Scrapped Joseph Pugsley c.1929 |
| 2 ft 0 in. gauge | | | | | | |
| | 0-4-0ST | Kerr, Stuart | 3090 | 1917 | ex-Balfour, Beatty | To Fry's, Keynsham contract |

## Charles Eckersley Daniel Portishead Docks Contract

| Name | Wheel arrangement | Builder | Works Number | Year | Source | Disposal |
|---|---|---|---|---|---|---|
| | 0-6-0ST | Manning, Wardle | 16 | 1860 | ex-Waterloo Main Colliery, Leeds | Sold out of service |
| Fareham | 0-6-0ST | Manning, Wardle | 51 | 1862 | ex-Somerset & Dorset Railway contract | Sold out of service |

## National Shipyard No. 3 Portbury

| Name | Wheel arrangement | Builder | Works Number | Year | Source | Disposal |
|---|---|---|---|---|---|---|
| | 0-4-0ST | Hunslet | 282 | 1882 | ex-Orrell Colliery, Wigan | To Tytherington Stone Co. |
| IWD No. 34 Portbury | 0-6-0ST | Avonside | 1764 | 1917 | New | To PBA (Inland Waterways & Docks) |
| | 0-6-0ST | Manning, Wardle Rebuilt Avonside | 726 / 7959 | 1879 / 1912 | ex-Rosyth Dockyard | Sold or scrapped |

## Canon's Marsh Gasworks (Bristol Gas Co. until 1st May, 1949)

| Name | Wheel arrangement | Builder | Works Number | Year | Source | Disposal |
|---|---|---|---|---|---|---|
| 3 Fenwick | 0-4-0ST | Peckett | 1211 | 1911 | New | Scrapped c.July 196 |
| 1 J. W. S. Dix | 0-4-0ST | Hawthorn, Leslie | 2184 | 1891 | ex-Stapleton Rd Gasworks June 1951 | Returned Sept 1954 |
| | 4wDM | Ruston & Hornsby | 321731 | 1952 | New | To Gloucester Gasworks 1958 |

# Appendix Three

## Speed Restrictions

### Maximum Speed of Trains through Junctions and at Specified Places—continued.

#### FILTON JUNCTION TO DR. DAY'S BRIDGE JUNCTION.

| NAME OF PLACE. | DIRECTION OF TRAINS. | | Miles per Hour. |
|---|---|---|---|
| | From. | To. | |
| **PORTISHEAD BRANCH.** | | | |
| Portishead Branch | Speed at any point not to exceed | | 35 |
| Clifton Bridge | Bristol | Portishead | 10 |
| Oak Wood | At either end of Loop | | 20 |
| Pill | Bristol | Portishead | 10 |
| Portbury Shipyard | Single Line | Up Loop | 10 |
| Portbury Shipyard | Up Loop | Single Line | 10 |
| Portishead Gasworks | Portishead | Bristol | 15 |
| Bristol West Depot | West Depot | West Loop North Junction | 10 |
| (Goods Running Loop) | West Loop North Junction | West Depot | 10 |

| **BRISTOL HARBOUR LINE.** | | |
|---|---|---|
| Temple Meads to Wapping Wharf Junction Over Bathurst Basin Bascule Bridge | Either direction | 5 |

| **WAPPING WHARF JUNCTION TO ASHTON SWING BRIDGE.** | | |
|---|---|---|
| From Wapping Wharf Junction to New Cut side of Cumberland Road Bridge | Either direction | 5 |
| From New Cut side of Cumberland Road Bridge to Level Crossing opposite Underfall Yard | Either direction | 10 |
| From Level Crossing opposite Underfall Yard to Ashton Swing Bridge North Junction | Either direction | 5 |

| **ASHTON JUNCTION TO CANONS MARSH.** | | |
|---|---|---|
| Over Ashton Swing Bridge | Either direction | 5 |
| Over Junction Lock Swing Bridge | Either direction | 5 |
| Over all Level Crossings | Either direction | 5 |
| Any other point between Ashton Junction and Canons Marsh | Either direction | 10 |

Maximum Speed of Trains through Junctions and at Specified Places. GWR Working Time Table 1945.

A Railway Correspondence & Travel Society special headed by '8750' class 0-6-0PT No 9769 crosses Ashton swing bridge 26th September, 1959.  *Dr A. J. G. Dickens*

# Appendix Four

## Loads

### Branch Lines—continued.
#### BRISTOL DIVISION.

| SECTION | | CLASS OF ENGINE | | | | | | |
|---|---|---|---|---|---|---|---|---|
| From | To | 4073-4099 5000-5099 | 4003 to 4072‡ 100, 111-4016 4000-4032 4037 | 29XX 31XX 41XX 43XX 51XX 53XX 55XX 61XX | 49XX 59XX 63XX 69XX (‡except 4016, 4032, 4037.) | 63XX 64XX 73XX 78XX 81XX 83XX 93XX | 3306—3455 4400—4410 5500—5574 36XX 37XX 57XX 77XX 97XX | 3200—3219 3252—3291 2251—2296 | 0-6-2 T. "B" Group. | 0-6-0 and 0-6-0 T. | 0-6-2 T. "A" Group. | 2-4-0 T. Metro. 0-4-2 T. 48XX 58XX |
| | | Tons. | Tons. | Tons. | Tons. | Tons. | Tons. | Tons. | Tons. | Tons. | Tons. |
| Clifton Bridge | Portishead | — | — | — | — | — | 308 | — | 280 | — | 242 | 898 1334 |
| Portishead | Clifton Bridge | — | — | — | — | — | 308 | — | 280 | — | 242 | 900 1335 1336 |

Engine loads 1945.

Maximum Loads for Branch Freight Trains 16th September, 1957 to 8th June, 1958.

### Maximum Loads for Branch Freight Trains—continued

**B142**

| | | WORKING LOADS | MAXIMUM ENGINE LOADS | | | | | | | | | | | | |
|---|---|---|---|---|---|---|---|---|---|---|---|---|---|---|---|
| BRANCH | | Maximum number of wagons to be conveyed except by Trains specially provided for in the Service Books or by arrangement | For Group A Engines | | | For Group B Engines | | | For Group C Engines | | | For Group D Engines | | | For Group E Engines | |
| From | To | | Class 1 Traffic | Class 2 Traffic | Empties | Class 1 Traffic | Class 2 Traffic | Empties | Class 1 Traffic | Class 2 Traffic | Empties | Class 1 Traffic | Class 2 Traffic | Empties | Class 1 Traffic | Class 2 Traffic | Empties |

**PORTISHEAD AND BRISTOL**

| West Depot | Clifton Bridge | 46 | 40 | 48 | 60 | 46 | 55 | 69 | 80 | 50 | 60 | 80 | 66 | 79 | 100 | 80 | 96 | 100 |
| Clifton Bridge | Pill | 46½ | 27 | 32 | 41 | 31 | 37 | 47 | 62 | 34 | 41 | 62 | 41 | 54 | 72 | 44 | 53 | 88 |
| Portishead | Portbury South Sidings | 46½ | 45 | 54 | 60 | 52 | 62 | 78 | 80 | 56 | 67 | 80 | — | — | — | — | — | — |
| Portbury South Sidings | Pill | 45½ | 32 | 41 | 54 | 36 | 47 | 60 | 62 | 41 | 51 | 68 | — | — | — | — | — | — |
| Pill | Ashton Junction | 42½ | 36 | 45 | 54 | 42 | 50 | 63 | 80 | 45 | 54 | 85 | — | — | — | — | — | — |
| Ashton Junction | West Depot via West Depot Loop or Parson Street Junction | 60½ | 22 | 26 | 33 | 25 | 30 | 38 | 50 | 27 | 32 | 54 | 36 | 43 | — | — | — | — |

**ASSISTED TRAINS.**—See pages 135 and 137. The instructions contained herein do not in any way affect or remove the prohibition placed by the Chief Engineer on the working of certain types of engines over certain sections of line, although loadings may be given in the table for engines over portions of line which are prohibited for them.

**Note.**—Engines of the 2-6-2T 45XX type when working stopping Freight Trains between Yatton and Wells, can only convey the same maximum loads as shewn for Group A Engines, owing to the limited capacity of their water tanks. When running non-stop between these points, these engines may convey a through load of 22 Class 1 traffic or its equivalent. Engines classified "RED" in Group A, B and C must not work between Bristol and Witham and vice versa.

† Can be made up to 50 if not required to cross a train in the opposite direction at Congresbury or Cheddar respectively.
‡ When the "Block Section" is between Clifton Bridge and Portishead trains from Ashton Meadows to Portishead and vice versa may be made up to full engine load hauling 60 wagons.

LOADS 189

## BRANCH LINES

| SECTION | | CLASS OF STEAM ENGINE | | | | | | | |
|---|---|---|---|---|---|---|---|---|---|
| | | B.R. Std. 70XXX | B.R. Std. 92XXX 47XXX Maximum speed 60 m.p.h. | B.R. Std. 73XXX | B.R. Std. 75XXX 76XXX | B.R. Std. 82XXX | 32XXX Ex LMR Class 2* | B.R. Std. 78XXX | 14XX | Ex LMR Std. Class 6P* |
| | | 10XX 70XX Ex LMR Class 7P* | | 79XX 68XX Ex LMR Std. Class 5* | 78XX 73XX 41XX 56XX Ex LMR Class 4* | 55XX 36XX 94XX Ex LMR Class 3* | | Ex LMR Class 2 (465XX) 16XX 74XX | | |
| From | To | Tons | Tons | Tons | Tons | Tons | Tons | Tons | Tons | Tons |
| Parson Street Junction | Clifton Bridge | 455 | — | 420 | 420 | 392 | 364 | 336 | 220 | — |
| Clifton Bridge | Portishead | — | — | — | — | 308 | 280 | 280 | 242 | — |
| Portishead | Clifton Bridge | — | — | 420 | — | 308 | 280 | 280 | 242 | — |
| Clifton Bridge | Parson Street Jn. | 455 | — | — | 392 | 336 | 336 | 308 | 198 | — |

A—55XX only.   C—16XX only.   ‡—The load on the Branch will be subject to sufficient clearances between the crossover roads.
*—Between Mangotsfield and Bath Green Park only.

Engine loads for passenger, parcels, milk and fish traffic steam locomotives 17th June, 1963.

Engine loads for passengers, parcels milk and fish traffic diesel locomotives 117th June, 1963.

## TABLE 'E' continued—MAXIMUM LOADS FOR DIESEL LOCOMOTIVE HAULED PASSENGER, PARCELS, MILK AND FISH TRAINS ON BRANCH LINES

| SECTION | | CLASS OF LOCOMOTIVE | | | | | | | | | |
|---|---|---|---|---|---|---|---|---|---|---|---|
| | | D11– D147 | D6XX | D8XX | D1XXX | D15XX D16XX | D6300 –D6305 | D6306 –D6357 | D67XX D68XX D69XX | D7XXX | D95XX | D2XXX |
| From | To | Tons | Tons | Tons | Tons | Tons | Tons | Tons | Tons | Tons | Tons | Tons |
| Parson Street | Clifton Bridge | 550 † | 550 † | 550 † | 550 † | 550 † | 550 † | 550 † | 550 † | 550 † | 550 † | — |
| Clifton Bridge | Portishead | — | — | — | — | — | 550 † | 550 | — | — | — | — |
| Portishead | Clifton Bridge | — | — | — | — | — | 515 | 550 | — | — | — | 260 |
| Clifton Bridge | Parson Street | 550 † | 550 † | 550 † | 550 † | 550 † | 535 | 550 † | 550 † | 550 † | 550 † | — |

†—525 tons for Passenger Trains.

The loads quoted are the absolute maximum authorised for the various types of locomotive and are not in any way related to any timing schedules. Reference to the Regional and Divisional Locomotive Route Availability Notices and relative publications will give other appropriate restrictions, which must at all times be observed.

# Appendix Five

## List of Signal Boxes—continued.

| Distance Box to Box. | | NAME OF BOX. | TIMES DURING WHICH BOXES ARE OPEN. | | | | Whether provided with Switch. |
|---|---|---|---|---|---|---|---|
| | | | Week Days. | | Sundays. | | |
| | | | Mondays. | Other Days. | Closed at | Opened at | Closed at | |
| M. D | C. 65 | Ashton Junction ........... | 5. 0 a.m. | — | — | — | 10.20 p.m. H | No. |
| — | 34 | Ashton Swing Bridge South .. | 5.30 a.m. | — | — | — | } 2. 0 p.m. K | No. |
| — | 10 | Ashton Swing Bridge North .. | 5.30 a.m. | — | — | — | | No. |
| — | 7 | Avon Crescent ........... | 5.30 a.m. | — | — | — | | No. |
| — | 54 | Canons Marsh Crossing ..... | 5.40 a.m. | — | — | — | — | No. |
| F | 41 | Clifton Bridge ............ | 5. 0 a.m. | 5. 0 a.m. | L | 10.15 a.m. | After last train. | No. |
| 2 | 63 | Oak Wood ............... | 7.45 a.m. | 7.45 a.m. | 10.15 p.m. } | — | — | Yes. |
| 1 | 50 | Pill .................... | 7.45 a.m. | 7.45 a.m. | 10.15 p.m. } | — | — | Yes. |
| 1 | 14 | Shipyard Sidings .......... | 7.45 a.m. | 7.45 a.m. | 10.15 p.m. } | — | — | Yes. |
| 2 | 45 | Portishead ............... | 5. 0 a.m. | 5. 0 a.m. | After last train. | 10.15 a.m. | After last train. | No. |
| G1 | 44 | Congresbury ............ ⎫ | | | | | | } No. |
| 2 | 74 | Sandford and Banwell ..... | | | | | | } No. |
| 3 | 39 | Axbridge................ | | | | | | No. |
| 1 | 57 | Cheddar ................ | | | | | | No. |
| 4 | 21 | Lodge Hill .............. ⎬ | For first train. | For first train. | After last train. | During train service. | | No. |
| 2 | 37 | Wookey ................ | | | | | | No. |
| 1 | 4 | Wells, Tucker Street....... | | | | | | No. |
| — | 7 | Wells, S. and D.* | | | | | | Yes. |
| — | 0 | Wells, E. Somerset ....... | | | | | | No. |
| 4 | 68 | Shepton Mallet .......... | | | | | | No. |
| 3 | 28 | Cranmore .............. ⎭ | | | | | | No. |
| 5 | 47 | Witham ................. | — | Open | continuously. | — | — | No. |
| E3 | 29 | Clevedon ................. | — | Closed. | — | (Worked as | Ground frame | } No. |
| — | — | Berkeley Loop Junction .... | 7. 0 p.m. | 7. 0 p.m. | 2. 0 a.m. | — | — | Yes. |
| 1 | 2 | Berkeley Station .......... | 7. 0 a.m. | 7. 0 a.m. | 8. 0 p.m. | — | — | West. |
| 1 | 15 | Sharpness South ......... | 4.30 a.m. | — | — | — | 4.30 a.m. | No. |
| — | 55 | Sharpness Station ........ | 4.30 a.m. | — | — | — | 4.30 a.m. | No. |
| 1 | 26 | Severn Bridge ............ | 4.45 a.m. | 4.45 a.m. | 11.30 p.m. | — | --- | No. |
| — | — | Severn Bridge Station ..... | 4.40 a.m. | 4.40 a.m. | 11.30 p.m. | — | — | No. |
| 2 | 30 | Otters Pool.............. | 4.20 a.m. | 4.20 a.m. | 11.30 p.m. | — | — | No. |
| 3 | 37 | Faringdon ............... | — | — | — | — | — | — |
| 3 | 45 | Malmesbury ............. | | | | | | |
| 2 | 10‡ | Kingsdown Road ......... | For first train. | For first train. | After last train. | As required. | As required. | Yes. |
| 3 | 38 | Highworth Station ....... | | | | | | No. |

**D**—From Parson Street. **E**—From Yatton West Junction. **F**—From Ashton Junction. **G**—From Yatton West. **H**—Or after passing of 9.48 p.m. Portishead. **K**—Or as ordered by Control. * Wells S. and D. Box is switched out after last S. and D. train on Saturdays until 6.30 a.m. Mondays, also when any G.W. early or late train is run on other days. ‡—From Highworth Junction.
**L**—Closed after 12.20 a.m. Goods from Portishead has cleared.

## WORKING OF ENGINES IN STEAM COUPLED TOGETHER.

### MAIN LINE ROUTES.

Instructions relative to the double heading of engines in steam on main line routes, and also the types of engines permitted to assist the 4-6-0 60XX "King" class, are given on page 142 of the General Appendix to the Book of Rules and Regulations, dated August 1st, 1936.

### ROUTES OTHER THAN MAIN LINES.

On the sections of the G.W. system not dealt with in the Main Line route instructions, not more than two engines in steam of the appropriate types in the group or groups authorised to work, may be coupled together and worked at customary speeds, except in those cases where special regulations are laid down to govern the working of engines in steam coupled on the section of line concerned. These cases form the subject of local instructions, and the speed limitations, where specified, must be strictly adhered to.

The " double heading " or " assistance " in the foregoing is subject to the following special regulations :—

### SWINDON SHED AND SWINDON FACTORY

Subject to the speed not exceeding 20 m.p.h., any group of not more than three engines (with the exception of the 60XX " King " class—singly or separately), may be worked coupled together between these points.

### BRISTOL (MARSH JUNCTION) TO FROME VIA RADSTOCK SECTION.

Tender engines authorised to work over this section may work double-headed funnel towards tender. Tank engines, as below, may work double-headed over the River Avon Bridge at 22 m. 70 chs. provided the speed does not exceed 15 miles per hour:

Two " Yellow " engines.
A " Yellow " and " Uncoloured " engine.
Two " Uncoloured " engines.

Other combinations are prohibited.

### SEVERN BRIDGE. SEVERN AND WYE JOINT LINE.

Two engines coupled together must not in any circumstances be run over the Severn Bridge (S. and W. Joint Line.)

Signal Boxes 1945.

# Appendix Six

## Log of Runs

3rd June, 1963 (Whit Monday)

Loco: 'Ivatt' Class 2 2-6-2T No. 41207
Coaches: 3 Southern Region
Average speed 27 mph

| Time Min Sec | Distance Miles | Station | Actual Arrival h m s | Scheduled Arrival h m s | Actual Departure h m s | Scheduled Departure |
|---|---|---|---|---|---|---|
| - - | - | Bristol (TM) | - - - | - - - | 4 35 45 | 4.35 |
| 3 15 | 1 | Bedminster | 4 39 0 | - - - | 4 39 18 | 4 38 |
| 2 27 | 1¾ | Parson Street | 4 41 45 | - - - | 4 49 20* | 4 41 |
| 2 52 | 3 | Ashton Gate | 4 52 12 | - - - | 4 52 18 | 4 44 |
| 1 39 | 3½ | Clifton Bridge | 4 53 57 | - - - | 4 54 29 | 4 46 |
| 6 59 | 7 | Ham Green Halt | 5 1 28 | - - - | 5 1 47 | 4 53 |
| 2 20 | 7¾ | Pill | 5 4 7 | - - - | 5 4 28 | 4 55 |
| 6 6 | 11½ | Portishead | 5 10 34 | 5 5 | | |

Total running time 25 minutes 38 seconds
*Adverse signals

Loco: 'Ivatt' Class 2 2-6-2T No. 41207
Coaches 3 Southern Region through to Bournemouth West.
(Full and with standing passengers)
Average speed: 27.6 mph

| Time Min Sec | Distance Miles | Station | Actual Arrival h m s | Scheduled Arrival h m s | Actual Departure h m s | Scheduled Departure |
|---|---|---|---|---|---|---|
| - | - | Portishead | - - - | - - - | 5 17 18 | 5 15 |
| 6 12 | 3¾ | Pill | 5 24 30 | - - - | 5 27 50 | 5 25 |
|  |  | Ham Green Halt |  |  |  |  |
| 8 18 | 8 | Clifton Bridge | 5 36 8 | - - - | 5 37 35 | 5 34 |
| 1 15 | 8½ | Ashton Gate | 5 38 50 | - - - | 5 39 40 | 5 36 |
| 3 15 | 9¾ | Parson Street | 5 42 55 | - - - | 5 44 10 | 5 40 |
| 2 20 | 10½ | Bedminster | 5 46 30 | - - - | 5 47 21 | 5 43 |
| 3 43 | 11½ | Bristol (TM) | 5 51 4 | 5 46 | | |

Total running time 25 minutes 3 seconds

TM – Temple Meads

# INDEX

Accidents 35, 40, 45, 54, 146, 178
Acts of Parliament 7, 8, 10, 27, 28, 32, 35, 45, 141, 146, 164, 169, 175
Albright & Wilson 48, 49, 103, 119, 185
Anglo-Saxon Petroleum Co 39
Ashton Containers 45, 65, 78, 185
Ashton Gate 33, 47, 52, 54, 65 *et seq.*, 116, 126, 129, 134, 140, 142, 161
Ashton Saw Mills 45, 65
Atmospheric Railway 7
AvonMetro 54
Avonmouth 5, 10, 35, 36, 42, 50, 119, 126

Barnet & Gale 27, 30
Barry, W. 36
Batchelor, Mr 13
Bedminster 8, 46, 54, 56, 77
Board of Trade 15, 36, 131, 145
Bower Ashton 133, 142, 164, 166, 168
BRIDGE
  Ashton Swing 46, 119, 147, 158, 160, 162 *et seq.*, 166, 187
  Bathurst 143, 145, 146, 154, 156
  Clifton Suspension 7, 9, 10
Bristol & Exeter Railway 8, 12 *et seq.*, 24, 36, 65, 95, 123, 143
Bristol City Football 65, 67, 119
Bristol Corporation 27, 35. 36, 63, 95, 141, 143, 162 *et seq.*, 178
Bristol Harbour Line 46, 65, 132, 143 *et seq.*, 151 *et seq.*
Bristol Port Railway & Pier 10, 27, 81
Bristol Tramways & Carriage Co 44, 51, 52
British Association 32
British Petroleum 39, 96
Brunel, I. K. 7
Budleigh Salterton 48
Butler, A. 88, 90
Butler, W. 175
Byers, S. 137

Campbell, P. & A. 36
Canning, T. 9
Canon's Marsh 65, 119, 121, 132, 143 *et seq.*, 162, 163, 169 *et seq.*, 177 *et seq.*
Cardiff 18, 23, 123, 126
Cardiff & Portishead Steamship Co 22
Castle M. 9, 23
Cavenagh, H. E. B. 105
Central Electricity Generating Board 49, 52
Central Stores Depot 44
Clevedon 40, 44
Clifton Bridge (station) 12 *et seq.*, 32, 36, 40, 42, 44, 45, 47, 49, 52, 58, 71 *et seq.*, 83, 86, 112, 119, 121, 123, 131, 132
Cowlin W. 112
Crawshay, Bailey & Co 13
Croome, Mr 8
Crossley, Flying Officer 49
Culverhouse P. E. 46, 175
Curry, Mr 11, 14

Daniel, C. E. 30
Daniel, J. F. R. 9, 24, 32
DOCK
  Portishead 9, 17, 24, 27 *et seq.*, 32, 35, 36, 39, 49, 77, 78
  Royal Portbury 6, 44, 53, 54, 89, 90, 121, 132 *et seq.* 136 *et seq.*
  Rownham 11
Drysdale, Mr 11, 13
Dublin 124
Durnford & Son 58

Esso 49
Evans D. 112

Ford, J. 31
Foss, Mr & Mrs 112
Fry, L. 31
Fry, R. 9

Fudge, R. 9
Furness Railway 9, 73

Gardiner, Dr 154
Giles & Son 61, 65
Godsell, Mr 11, 14
Gordano School 114, 115
Grace, Mr 7
Gradients 12, 13
Great Western Railway 7, 18, 22, 24, 31, 35, 36, 40, 103, 143, 169, 172

Ham Green 12, 46, 83 *et seq.*, 128
Hill, H.155
Hotwells 10

Ilfracombe 18, 22, 23, 103
Imperial Tobacco Co 45, 60, 148, 153

Keys, Messrs 39

Leigh Woods 10, 12, 13, 17, 81
Locomotives 41, 42, 45, 105, 117 *et seq.*, 137, 138, 148, 149, 175, 176
Lyn Steamship Co 10, 23
Lyon, R. 32
Lynmouth22
Lysaght, J. 163

MacPherson, Mr 11, 14
Mardon, Son & Hall 154
Marples, E. 52
Maudsley, Messrs 28
May & Hassell Ltd 60
McAlpine A. 133
McClean J. R. 8, 9, 11, 17, 27
McNeil, Sir J. K. 8
Midland Railway 18, 23, 24, 36, 42, 103, 169
Midland & South Western Junction Railway 9
Miles, Sir W. 13
Ministry of Fuel & Power 49
Mitchell, R. 112
Moore, J. 7
Mustad, Messrs 103
Mylne, W. C. 7

National Shipyard Co 12
Newport 18, 123, 137
Nightingale Valley 11, 79, 80, 84
Northville Building 60

Oak Wood 46, 71, 73, 82 *et seq.*, 121, 132

Paddington 24
Parson Street 46, 47, 54, 57 *et seq.*, 71, 133, 134, 137, 146, 148
Permanent Way 15, 58, 146
Pier, Portishead 8, 15, 17, 23 *et seq.*, 35, 36, 103
Pill 6, 14, 15, 40, 42, 46, 47, 54, 73, 77, 83 *et seq.*, 121, 126, 131 *et seq.*, 137, 138, 140, 142
Plimsoll, S. 154
Port of Bristol Authority 41, 42, 48, 137, 149, 184
Portbury 7, 15, 42 *et seq.*, 46, 52, 54, 93, 95, 116, 141
Portbury Pier & Railway Co 7, 19 *et seq.*
Portbury Shipyard 43, 84, 90 *et seq.*, 121, 126, 132, 149
Portishead 5 *et seq.*, 14, 15, 17, 19 *et seq.*, 30, 40 *et seq.*, 52 *et seq.*, 58, 63, 73, 77, 86, 74, 85, 117*et seq.*, 126, 130, 131, 140 *et seq.*
Portishead Junction 36, 41, 46, 58, 60
Power stations 45 *et seq.*, 53, 77, 99, 103, 105
Proctor, H. & T. 154

QUARRIES
  Abbot's Leigh 81
  Black Rock 39, 40, 103

Conygar 39
Corporation 62, 64, 81
Greenland 36
Hickory 62, 64, 81
Netham 61, 62, 64, 81
Wethered & Worsley 81
Whitestone 81

Redcliffe goods yard 148, 150, 151, 153, 154
Rich Col F. H. 36
Richardson, C. 143
River Avon 5, 10, 36, 88
Robertson, S. 175
Robinson, R. 9
Rochester 41
Rownham 11, 13, 73, 77
Rudders & Payne 65
Rumsey Mr. 7

St Anne's Board Mills 53
Severn Kraft Mills 46, 95
Shell, Messrs 42
Sheppard, P. 112
Shopland, E. H. 95
Signalling 31, 36, 39, 42, 46, 58, 60, 65, 73, 74, 77, 83, 84, 86, 88, 91, 112, 113, 131 *et seq.*, 148, 153, 155, 163, 164
South Devon Railway 7
South Eastern & Chatham Railway 41
South Staffordshire Railway 9
SHIPS
  *SS Astro* 24
  *SS Magdeburg* 30
  *PS Bee* 24
  *PS Dart* 22
  *PS Ely* 22 *et seq.*
  *PS Gael* 24
  *PS Lyn* 24, 30
  *PS Taff* 22 *et seq.*
Stileman, F. C. 10, 11, 17, 22, 27, 28
Strike, 1911 40
Strachen & Henshaw 65

Taff Vale Railway 9
Temple Meads 22 *et seq.*, 46, 47, 55, 77, 78, 123, 126, 129, 130, 138, 142, 148, 150, 151, 155, 172
Tenby 24
Toluol 17
Toogood F. W. 65
Tredwell, W. 11, 12, 14, 22
TUNNEL
  Clifton No 1 10, 11, 14, 16, 54, 77, 79
  Clifton No 2 16, 36, 81
  Clifton No 3 16, 36
  Pill 16, 84 *et seq.*, 135, 137
  Redcliffe 154, 155, 166
  Sandstone 81, 83

United Alkali Co 36

VIADUCT
  Chapel Pill 83
  Pill 14, 84, 134 *et seq.*, 140
  Portishead 14

Walker & Burges 9
Wapping 46, 54, 119, 121, 132, 141, 144 *et seq.*, 148, 149, 155, 157, 158, 162, 163
Weatherley F. 9
Weston, Clevedon & Portishead Light Railway 9, 39, 40, 63, 103
West Depot 44, 46, 60, 130, 172
Wills, Sir G. 81
Woodward, G. R. 9
WORLD WAR
  First 5, 40, 42, 65, 77, 91
  Second 5, 40, 47, 71, 78, 79, 84, 117, 146, 166, 172, 175, 178

Yolland, Col 15, 145, 146